Farrell - another
text on Ed. Admin

Ecosophy and Educational Research for the Anthropocene

Problematizing the aims of education in the Anthropocene, this text illustrates the value of relational psychoanalytic theory in the study and practice of education amidst the climate crisis.

Illustrating how dominant educational theory fails to acknowledge climate precarity and the consequences of living beyond the Earth's carrying capacity, *Ecosophy and Educational Research for the Anthropocene* calls for a reorientation of scholarship to decentre the human subject. The author discusses the evolution of intersubjective psychoanalysis to make a case for a turn to relational and psychoanalytically informed educational research. Chapters foreground areas for educational researchers to consider in pursuing intersubjective inquiries into the affective dimensions of curriculum and pedagogy to foster an emergence of eco-attunement and ecosophical educational research (EER).

By framing an ecosophical approach, this book enables educational leaders, researchers and educators to fulfil their responsibility to engage in educational praxis which is contextually responsive, relationally attuned and recognizant that we cannot be studied apart from our connections to the planet.

Alysha J. Farrell is Associate Professor of Education at Brandon University, Canada.

Routledge Research in Education, Society and the Anthropocene

This series offers a global platform to engage scholars in continuous academic debate on key challenges and the latest thinking on education and society and the role of education in the Anthropocene. It provides a forum for established and emerging scholars to discuss the latest debates, issues, research and theory across the field of education research relating to society and the Anthropocene.

Rethinking Education in Light of Global Challenges
Scandinavian Perspectives on Culture, Society and the Anthropocene
Edited by Karen Bjerg Petersen, Kerstin von Brömssen, Gro Hellesdatter Jacobsen, Jesper Garsdal, Michael Paulsen and Oleg Koefoed

Ecosophy and Educational Research for the Anthropocene
Rethinking Research through Relational Psychoanalytic Approaches
Alysha J. Farrell

Justice and Equity in Climate Change Education
Exploring Social and Ethical Dimensions of Environmental Education
Edited by Elizabeth M. Walsh

Ecosophy and Educational Research for the Anthropocene
Rethinking Research through Relational Psychoanalytic Approaches

Alysha J. Farrell

Ecosophy is a philosophy of ecological harmony & equilibrium.

Anthropocene denotes the current geological era of humans.

First published 2022
by Routledge
605 Third Avenue, New York, NY 10158

and by Routledge
2 Park Square, Milton Park, Abingdon, Oxon, OX14 4RN

Routledge is an imprint of the Taylor & Francis Group, an informa business

© 2022 Taylor & Francis

The right of Alysha J. Farrell to be identified as author of this work has been asserted by her in accordance with sections 77 and 78 of the Copyright, Designs and Patents Act 1988.

All rights reserved. No part of this book may be reprinted or reproduced or utilised in any form or by any electronic, mechanical, or other means, now known or hereafter invented, including photocopying and recording, or in any information storage or retrieval system, without permission in writing from the publishers.

Trademark notice: Product or corporate names may be trademarks or registered trademarks, and are used only for identification and explanation without intent to infringe.

Library of Congress Cataloging-in-Publication Data
Names: Farrell, Alysha J., author.
Title: Ecosophy and educational research for the anthropocene : rethinking research through relational psychoanalytic approaches / Alysha J. Farrell.
Description: New York, NY : Routledge, 2022. | Series: Routledge research in education, society and the anthropocene | Includes bibliographical references and index.
Identifiers: LCCN 2021031129 (print) | LCCN 2021031130 (ebook) | ISBN 9780367457099 (hardback) | ISBN 9781032146133 (paperback) | ISBN 9781003024873 (ebook)
Subjects: LCSH: Psychoanalysis and education. | Environmental education. | Education–Research.
Classification: LCC LB1092 .F37 2022 (print) | LCC LB1092 (ebook) | DDC 370.15–dc23
LC record available at https://lccn.loc.gov/2021031129
LC ebook record available at https://lccn.loc.gov/2021031130

ISBN: 978-0-367-45709-9 (hbk)
ISBN: 978-1-032-14613-3 (pbk)
ISBN: 978-1-003-02487-3 (ebk)

DOI: 10.4324/9781003024873

Typeset in Sabon
by KnowledgeWorks Global Ltd.

To Vanya & Kirill,
to all who are feeling unsettled by the
changing Land, Water and Skies,
to my-more-than human relatives,
and to generations of school children yet to
come

Contents

List of Figures ix
Acknowledgements x

What Does It Mean to Educate in a World That is Prepared to Go on Without Us? 1

1 Renovating the Schoolhouse When the World is on Fire 16

2 Education and the Visuality of the Anthropocene 33

3 A Prairie Elegy for the Discerning Consumer 47

4 The Emotional Impact of Living in the Climate Crisis 58

5 Anthropocentric Fantasies Entangled in the Pain of Eco-Grief 73

6 Relational and Psychoanalytically Informed Education in the Anthropocene 85

7 Cultivating Solidarity as the Climate Crisis Intensifies 100

8 What We Can Learn from the COVID-19 Pandemic 119

9 Monstrous Feelings and Ghostly Echoes of the Anthropocene 132

10 Ecosophical Educational Research on the Edge of the Anthropocene 145

11 Ecosophical Educational Leadership	159
12 Teacher Education on the Edge of the Anthropocene	172
Get Busy Living	189
Appendix	194
Index	197

List of Figures

2.1	A Pair of Geese in Flight	33
2.2	Gobbler Claims Space	34
2.3	Foraging for Fish	35
2.4	Cementing a Bond	35
2.5	Swirl of Carp at Twilight	39
2.6	The Mise-En-Scène of the Anthropocene	45
3.1	The Weedy Tangles of Akiya	48
3.2	Beak Wars	49
3.3	Harry is Hollow	50
3.4	Zombie Hot Wheels	51
3.5	The Fragility of Butterfly Wings	52
3.6	Mother and Calf Share a Secret in the Pasture	53
3.7	You Aint Nothin But a Slime Ball, Cryin All the Time	54
3.8	Digital Tethers	55
3.9	Taming Meme-i-mals	56
3.10	When Smoke Gets Under Your Eyelids	56
12.1	A View from the Arlington Street Bridge	173
12.2	A Pigeon Soars	175

Acknowledgements

A relational home is one of gratitude. My appreciation and acknowledgements flow through the affective, ecological and cultural registers of my life.

The geese, terns, willow trees, and the lake breeze around our cottage at Delta Beach, have nourished me in countless ways. The writing has sometimes felt like a solitary process, but in other ways, it has been a complete immersion in the wider web of relations – an opportunity to connect more deeply with family, friends, colleagues and my more-than-human relatives.

This book draws on the work of relational psychoanalysts, climate psychologists, environmental activists, artists and journalists. Without their courageous work and insights, this book would not exist. I am deeply grateful for all that I have learned.

In the winter term of 2021, I taught an Innovative Pedagogies grad course. The students and I explored land-based pedagogies and storytelling in relation to the climate crisis. Their thoughtful engagement with the readings, artworks and creative activities, continues to inspire me.

I'd like to thank my colleagues and friends at Brandon University who have generously shared their time and wisdom with me. Our conversations have nourished this work in countless ways.

Thank you to Ellie Wright, AnnaMary Goodall and the rest of the Routledge team for helping me to share these ideas with others.

To Ollie and Maiv, thank you for insisting at least three times a day it was time to play fetch even when my facial expressions might have conveyed it was the last thing I wanted to do. The book would have been much harder to write in the absence of your wagging tails.

Thanks, too, to my son's Vanya and Kirill, for being such amazing humans. You finished high school and started university classes in the middle of a pandemic. I am so proud of you, especially because you are so kind to Others.

Above all, thank you to my husband, Trent Sloane, who has always encouraged me to write. You made supper for the kids when I was busy finishing a chapter and you carried half the camera equipment on our walks through Riding Mountain Park. For these things, and the million other things you do to care for me, Vanya, Kirill, Maiv and Ollie, I am grateful.

What Does It Mean to Educate in a World That is Prepared [or must?] to Go on Without Us?

As educators, we walk hand in hand with students toward the future. We hold on tight enough to protect the young from the world and then we learn to let go so they can renew the world in ways we cannot possibly imagine. Yet, when it comes to the ecological crisis, we evade knowing climate precarity threatens to destroy the world in which we practice our beloved profession. Emily Dickinson (1896) reminds us in her poem, *Ghosts*, "One need not be a chamber to be haunted". *Ecosophy and Educational Research for the Anthropocene: Rethinking Research through Relational Psychoanalytic Approaches*, is an attempt to reconcile the disavowal of ecological collapse in a field that is deeply invested in preparing young people to renew the world for generations yet to come.

After 12 000 years of relative stability, we have entered the epoch of the *Anthropocene*. Anthropocene is a term popularized by paleoecologist Eugene Stoermer and atmospheric chemist Paul Crutzen in 2000, to acknowledge humans are the first species to exert a planetary influence on the Earth's climate and ecosystems. As of March 2021, the International Union of Geological Sciences had not yet approved the term as a recognized subdivision of geologic time. Despite its unofficial status, the term is used frequently in informal scientific contexts, and it has ignited several provocative inquiries about the end of nature, extinction and the climate emergency in the fields of education, art, philosophy and literature. The use of the term, "Anthropocene" is not without its critics. According to the United Nations Environment Programme (2020), the G20, an informal group of 19 countries, the European Union, finance ministers and central bank governors, account for 70% of greenhouse gas emissions. Most people are not to blame for the predicament we find ourselves in today – "perhaps technocene or plutocene or capitalocene would all be truer terms, since these gigantic globe-reshaping effects are caused by technology and overconsumption by the wealthy, and by capitalism" (Bhalla, 2021, para. 8). For other critics, the use of the term reinforces anthropocentrism at a time when many people are troubling the human-centric pillars in their fields. While I acknowledge the

[induced by capitalism, not individuals - saying other we other i't, i't showing i't]

important and thoughtful criticisms, I believe the term opens space in the field of educational research to decentre the ontological positioning of humans and illuminate the interconnectivity of all Beings.

Inequality is a defining feature of the Anthropocene and it will increase dramatically over the next three decades. Fledgling democracies will struggle as strongman leaders amplify us-versus-them talking points. The public discourse will coarsen, contributing to an upsurge in acts of racism, misogyny and violence. To placate younger voters, politicians will continue to greenwash policies and construct climate change as a creeping crisis (Boin et al., 2017), spinning the fantasy, we have infinite time to draw down emissions. A lack of felt urgency will make it harder for eco-coalitions to mitigate the acceleration of the sixth mass extinction. But there is a reckoning with reality on the horizon. We are at the back end of an ontological rupture. To face this difficult knowledge is to pry open our eyes and find ourselves looking over the precipice. Illusions of anthropocentric dominance, the hubris in thinking we just need to give people more facts and their behaviour will change, will not result in the profound shift we need to make if our species is to survive.

In Chapter 1, I discuss the real and metaphorical fires in our current context, the ones making it difficult to reorient education and educational research towards ecosophical ends. I begin with the boil water advisories in rural and remote First Nations communities in Canada and then address the Flint, Michigan water crisis to describe an erosion of the spirit of what the Lakota refer to as Mni wiconi (water is life) by Mad Max logic. Then, I explore how the brutality of late stage capitalism leads to educational policy and school renewal efforts that are antithetical to an attunement to the more-than-human world. I consider how the digital echo location process on social media shapes climate change discourse for better and for worse.

Using a feeling-photography, in Chapter 2, I pull attention to climate precarity and other planetary boundaries exceeded by human activity, to interrupt the pattern of ecological erasure in educational research. I describe some of the impediments restricting the field's visuality and then consider how photography, and narrative photography in particular, can illuminate the aesthetics of the Anthropocene. Moving against the dominant selfie-culture, I turn my lens towards Others to open a curriculum trajectory across different spaces and among different species. I evoke a feeling-photography as a way to, literally and figuratively, consider the values and the emotionality embodied in particular frames of curriculum and pedagogy that have been inherited and internalized.

To crawl under anthropocentrism in Chapter 3, I present 10 digital photos and elongated captions from a larger visual arts project I am working on called, *A Prairie Elegy for the Discerning Consumer*. The work is intended to open an interpretive space to examine the roots of the ecological crisis enmeshed in the field of educational research. More specifically, the work is an interrogation of the *leakiness of social necrosis*. The activities surrounding this wild and wilding inquiry, oscillate between walking outside, staging concepts in photographs, research, writing, sculpting objects found outside (within in a do no harm ethos), observing other creatures, daydreaming and noting compelling images in the aftermath, and re/configuring the material world into megabytes. The images represent the connections I have made among the mental, political and affective registers of late stage capitalism.

Education in the Anthropocene

Young people are increasingly aware of the precarity of our situation. Propelled by concerns for their future, and angered by the inaction of adults, students across the globe continue to walk out of school on Fridays to participate in climate strikes. Yet, in the field of education, we have yet to respond in any significant way to the harm the climate crisis will inflict upon the people we engage with in our classrooms every day. What is even more disheartening, is when young people critique the economic system (*"Okay, Boomer!"*) which fuels the massive losses in biodiversity, they are often met with cynicism or placations from adults. The ambivalence in the field runs counter to the socialization of educators who care deeply for other people's children. Teaching in the Anthropocene, then, thrusts the legal, ethical, moral and intergenerational obligations of in loco parentis into question.

With the amplification of authoritarian leadership discourses in many countries across the world, and the ubiquity of a politically territorial social media habitus, the question of how to engage democratically within and beyond educational contexts has become urgent. Alternative assemblages of subjectivity (Hardt & Negri, 2017) are needed to develop artful ecosophical praxis in educative spaces to intervene in group dynamics that entrench asymmetries of power. Education must accept its responsibility to prepare eco-conscious citizens, compassionate, relational people, who will live in harmony within the more-than-human world. Educational researchers, then, have a role to play in studying and disseminating knowledge about attuning to our more-than-human relatives, building a relational home for students, amplifying learning processes that privilege interdependence over individualism, increasing understanding about the affective dimensions of the climate crisis and fostering resiliency processes. This ecosophical reorientation will require a reconciliation of the field's own ambivalence about the danger we collectively face.

Wynes and Nicholas (2017) analyzed the coverage of climate change in high school curricula across Canada according to six core topics. In

Canada, there are three territories and 10 provinces, and each jurisdiction controls its own school system. The researchers found the learning objectives across the provinces and territories tended to focus on physical climate mechanisms, observable increases in temperature and anthropogenic causes with little focus on scientific consensus about climate change, the negative consequences of global warming or how humans could mitigate the worst effects of global warming (p. 1). It is no surprise, then, that despite the overwhelming scientific consensus that global warming poses serious risks to all ecosystems on Earth, many young people in Canada are studying in classrooms and lecture halls in which the urgency of the climate emergency is eclipsed by other disciplinary concerns. To further complicate matters, ambivalence about climate change has become even more virulent in Canada as right-wing politicians have deferred to the merchants of doubt in the fossil fuel industry.

One way to make a resonant sound in the silence is to use the terms climate crisis, ecological emergency or global warming instead of the term climate change within the context of transdisciplinary spaces of curriculum and pedagogy. Climate change is nebulous. It leaves room to debate the existence of global warming if your community's weather disaster has yet to make the national news. *Doesn't the climate change over time? There isn't 100% agreement yet in the scientific community. How bad could it really be?* In the province of Manitoba, the supplementary material included with the grade 11 chemistry curricula, under the unfortunate heading, *Climate Science Debate* says,

> "It is important for students to conduct research that is fair and representative of alternative, viable, scientific viewpoints on such a vital issue. Students should research climate science, as articulated by organizations such as The Friends of Science – Providing Insight into Climate Science. This large, international community of climate scientists, for instance, holds views quite contrary to what has been supported by Environment Canada and the United Nations' (UN) Intergovernmental Panel on Climate Change (IPCC) over the past decade".
>
> (Manitoba Education, 2006, p. 159)

It is important to note, Manitoba is not the only provincial educational authority in Canada to frame anthropogenic climate change as a question yet to be answered. Climate science curricula in many Canadian secondary schools gives little emphasis to scientific consensus, the impacts or the actions that people can take to mitigate the negative impacts of global warming (Wynes & Nicholas, 2017). Educational researchers could make a productive ruckus about such problematic omissions and evasions.

The Feelings of the Anthropocene

Climate conscious young people have been placed in a disorienting psychological position. Every day, their social media feeds fill with dire images of severe weather events, habitat losses and expanding endangered species lists. Some become overwhelmed, enraged, despondent or even numb. Then they go to school and take the same subjects, in the same order with the same (minimal) emphasis on global warming in much the same way as their parents and grandparents did. Young people are expected to sit at their desks as if everything is normal when the world feels anything but normal. When adults minimize visible threats to the survival of the human species it is and will be a dehumanizing force in the psychosocial landscapes of young people (Cunsolo et al., 2020; Hickman, 2019; Hickman, 2020). To put it another way, it is intergenerational malpractice for educators to ignore the apocalyptic words and images that increasingly roil in students' bodies as they metabolize the climate crisis.

Although the affective dimensions of learning have traditionally been the domain of those in educational psychology, the Anthropocene calls into question attempts to separate learning from its emotional charge. Monstrous feelings about the climate crisis are ever present, and to deny their existence, or to save them for the guidance counsellor's office when they appear as other symptoms, dislocates the emotional register from educative encounters. To address the increase in frequency and intensity of the emotional impacts of the climate crisis, we need to talk openly about what reinforces the silence of teachers, leaders and policy makers, who one hand, profess a genuine love for other people's children and on the other, remain rooted in a system that disavows their intergenerational responsibility to protect the lives of children in the face of harm. Inquiring into the silence requires relational psychoanalytic concepts such as ambivalence, disavowal and denial to be introduced into professional conversations about the affective dimensions of climate precarity.

The failure to respond to the threats posed by global warming is often constructed as a problem with having the right information in education, that is, if we just get the correct information to the right people, they will respond. If this were the case, we might imagine climate scientists and environmental activists, beating down the doors of schools and universities to share their Power Point presentations in assembly halls. Others believe big tech, Twitter, Facebook and Instagram, are the moons shifting the tide of misinformation contaminating the public discourse. Although social media bubbles are problematic, and it is true algorithms reinforce the confirmation biases of digital users, the ubiquity of social media does not account for the number of young people and adults in the education system who remain unresponsive to climate change, a phenomenon that is everywhere and nowhere. In this paradoxical realm, we are haunted by something we refuse to see and cannot bear to imagine.

That said, climate ambivalence is not impossible to understand. For the most part, human beings are averse to the pain of loss, and speaking about climate change forces us to confront many losses. When the obligation of stewardship surfaces or when someone says your lifestyle will have to change to draw down emissions, most people will defend themselves against the discomfort. Others, might be ambivalent because they do not know if responding to climate change is more important than digging out from the economic hardships created by the COVID-19 pandemic. To open a space to discuss ambivalence about anthropocentrism, I build an ecosophical orientation at the intersection of corporeal vulnerability, eco-anxiety and grief.

In Chapter 4, I make the case educational research in the Anthropocene must be, in part, predicated on its ability to engage with *telluric emotions* because we cannot be healed or studied apart from the planet. Using a relational psychoanalytic frame, I interrogate how emotionally charged information about the impacts of global warming elicits defence mechanisms such as disavowal, projection and splitting in the field. To care for our own and the defensive responses of others, I call for a reorientation toward ecosophical research and robust inquiries into what it means for educators to be compassionate witnesses who are capable of creating holding environments for climate dread.

Recently, I became acquainted with the work of Marc Schlossman, the creator of a series of photographs called, *Extinction*. He photographed several specimens of extinct and endangered species in the zoology and botany collections of The Field Museum of Natural History in Chicago (Heinz, 2017). Each of his photographs is accompanied by a haunting commemoration of a specimen's existence and disappearance. One of the photos that left me sensory-struck was the image and description of the Xerces Blue butterfly. The caption includes a poetic tribute from veterinarian and author Mark Jerome Walters who laments, "When the tiny wings of the last Xerces Blue butterfly ceased to flutter, our world grew quieter by a whisper and duller by a hue" (Heinz, 2017). As I explored more of Schlossman's photographs, I began to think about his work as the reclamation of disenfranchised grief within a space of public pedagogy.

There are beautiful and life-affirming feelings entangled with the sources of our collective grief. We grieve what we love, and when we mourn, the process eventually makes room for joy, passion and hope. I believe education should open spaces of curriculum and pedagogy, to grieve what we losing. Taking the affective register as my starting point, the broader aim is to inspire new thinking about embodied vulnerability as curriculum within the more-than-human world. To create a holding environment for grief and loss in educational research in Chapter 5, I assert psychoanalytically informed educational research could lay open the endocryptic identifications in the aims, processes and practices that cohere the field to the figurative ghosts of late stage capitalism. I discuss why young people's disenfranchised eco-grief, and the laissez-faire attitude of adults, makes adults emotionally illegible to climate coconscious students. I explain why it is integral to acknowledge the trauma of ecocide and how important it is to give language to mourning before students engage in preservative repression. Using the *Flood of 97* in Manitoba, Marc Scholssman's photographic inquiry called, *Extinction*, and the graffiti artist Banksy's apocalyptic pop-up exhibition called *Dismaland*, I share why it is critically important to make room for moody pessimism in education.

Calling forth monstrous feelings to dislocate dominant educational theories from an idealized future through a discussion of visuality and narrative photography, I propose a dramatic reconceptualization of the aims of educational research. As I etch images of what it means to fully recognize the Other in the context of education. I want to resuscitate the category of the same (Ruti, 2018) in order to challenge all of us to accept the inherent responsibility in education to preserve children's lives in the face of environmental harm. While acknowledging the arduous and anxiety filled work ahead, I look over the edge of the Anthropocene to make meaning of education in relational psychoanalytic terms. I wonder, how might we conceive of ourselves as educational researchers when the habitability of the Earth is not a given anymore? How might a poetics of relationality help us to deal with the existential anxiety that perpetuates happy affects in the face of the climate crisis?

In Chapter 9, I explore the epistemology of ghosts haunting the Anthropocene. The haunting I refer to, is not the return of something dead, like contemporary calls to return to some romanticized image of nature, but a haunting presentism, one in which the collective embodied awareness of the climate crisis interferes with the day-to-day visuality of interpretation. Using the work of Abraham and Torok (1971; 1972), I explain how trauma remains hidden from a sufferer's conscious awareness through the process of incorporation. I argue the monster figures roaming our interiority can draw our attention to humanity's role in the unmaking of ecosystems, that is, *the process of becoming monster*, so we may better understand the legacies of our species' biomorphic powers. As distillations of what frames our attachments and fears in the context of creative work, monsters offer just the right dose of wry detachment and abjection to remind us we are made of strong things. In the spirit of inviting our monsters out to play, I share three arts-based methods to trouble human narcissism and techno-optimism in the field.

The Focus of the Work

I want to draw attention to the life-enhancing ideas education can bequeath, and how education can be a positive force in preparing children for an uncertain future. That said, the arguments contained in the book are not about greenwashing curricula or building more outdoor classrooms. No, I am calling for an eco-orientation of the aims, methods and relationality of education. My project contextualizes educational research in the Anthropocene as the search to live well with our more-than-human relatives. I join with other scholars in curriculum and pedagogy, educational leadership and psychoanalysis who engage in research that is contextually responsive, intersubjective and focused on living within the Earth's carrying capacity.

To disrupt anthropocentric teaching and learning, is to situate the foundational aims of education within the web of relations that supports life on this planet. Ecosophical education is context specific in terms of the questions, resources and strategies educators use in spaces of curriculum and pedagogy, but it is founded upon the universal values of protecting nature, love for self and Others, defining success in terms of wellbeing, fostering interdependence and climate citizenship. In Chapter 12, I explore what a teacher education program would look like if it took the intergenerational obligations of *in loco parentis* to heart in response to the climate crisis. I begin by framing an image of ecosophical teacher education in reference to a walk I took along the Arlington Street Bridge in Winnipeg a number of years ago. Using a couple of key lessons from the walk and the pigeon aesthetics on display that day, I make a relational psychoanalytic turn to breathe life into the emotions, dreams, interrupted words, silences and thought fragments infusing curriculum and pedagogy. Then, I draw a more detailed picture of relational teacher education, one in which teacher candidates develop confidence in living classroom inquires open to the unbidden. In this new evocation of ecosophical teacher education, I imagine teacher candidates working alongside mentors who weave a critical pedagogy of place and nurture a reverence for the diversity among all biological and cultural systems.

Ecosophy

The prefix *eco* in ecosophy, comes from the ancient Greek word *oikos* which refers to three related concepts, *the family*, *home* and by extension, *habitat*. The term ecosophy emerged in the work of French psychoanalyst Félix Guattri (2008) and Norwegian philosopher and environmentalist Arne Naess (2008) around the same time. Naess's ecosophy enlarges the sphere of Beings and objects with which people experience a kinship. His wager is that once people feel more connected to other objects and living Beings in the natural world, they will assume some responsibility to protect and preserve them. For Guattari, ecosophy is the complex interaction "between the three ecological registers (the environment, social relations and human subjectivity)" (p. 28). He views human and nonhumans as *actants*, meaning they can both operate as objects or subjects and embody aspects of subject–object relations at the same time. When I drink peppermint tea, the water in my mug is an object but I quickly become the object when a fast-moving current in the river sweeps me off my feet. When Alexa, the name given to Amazon's digital voice assistant, frustrates me because she plays the wrong song, it reveals something about both of us. She has more to learn to accurately read me through her natural language processing (NLP) system and I reread myself through the affective responses generated in our interactions. In my encounters with Alexa, I sense my frustration says something important about my own NLP enactments! The blurry lines of subjectivity show us identities are always in a state of flux in the Anthropocene and that our relationships are constantly shapeshifting into new assemblages.

Within the predatory capitalist system of oppression that infiltrates much of our existence, an attunement to environmental concerns must include more than the natural environment register. We must reach into the affective, cultural and existential territories to breach the illusory lines that separate the digital world from the (real) one, for example. Guattari's (2008) plurality of ecologies suggests that the environment is not the fishbowl that surrounds the goldfish. It is a dynamic network of relations that we define and that defines us in perpetuity. It means the definitive lines we arbitrarily draw to separate humans from their natural and digital worlds do not hold. Not only do the illusory demarcations make it more challenging to notice the dynamism of life, it circumvents the enactment of multi-vocal dialogical spaces needed to move towards transformation. It is the organic fusions and dissolutions of a multiplicity of connections among Beings that will produce the radical dialogical explorations needed to respond to the climate emergency.

An Ecosophical Orientation in the Field of Education

I have referred to the reimagination of education and educational research in the Anthropocene in ecosophical terms. Ecosophical education cannot be reduced to the addition of a new climate change unit in a science course nor can it be the greenwashing of old disciplinary questions. Managerial approaches to ecology, such as the ones framing many units in sustainability education, are insufficient because they make the brutality of capitalism less horrifying in appearance without challenging existing models of production and consumption.

Ecosophical education does not separate humans from their environment. It views each Being on the planet as an interaction. Citton (2014) draws the link between ecosophy and relationism when he describes an ecosophical perspective as one in which "…individuals do not pre-exist the relations that shape them" (cited in Antonioli, 2018, p. 2). Ecosophical education disrupts anthropocentrism and gives salience to animals and plants in spaces of curriculum and pedagogy. Further, it is committed to living in balance with our more-than-human relatives. It is grounded in an ethos of mutual recognition, dialogical fascination with the Other, non-violence, democratic participation and compassionate witnessing. Ecosophical education is affectively charged, and the attunement to telluric emotions *is education*. Pedagogy, then, is the art of becoming more attentive to students' emotions as they encounter new ideas and increasingly aware of what escapes students' awareness but still manages to influence their actions. Teachers are always at work in the realm of disparities, awkward silences and fleeting impressions while articulating partial meanings of felt experiences. I hold these tenets close to my heart whenever I make judgments about the goodness and significance of education and educational research.

One of the criticisms of an ecosophical approach to the theory and practice of education, is that it is resistant to application (Greenhalgh-Spencer, 2014). I interpret ecosophy's resistance to application as a welcome reprieve from the technocratic approaches and rubric fetishizing that goes on in many school systems right now. I also see the tension as a source of energy to crack open spaces of curriculum and pedagogy that run counter to the predatory capitalist assemblages that infect the mental, affective, and ecological registers in our lives. Having said that, throughout the book, I do include some images of praxis to enliven ecosophical concepts within the context of schooling and research. They are not instructions or recipes, they are in my mind, an exercise in pedagogic echolocation, a way to connect with Others who are also exploring ecosophical trajectories in the field.

In August, 2019, Greta Thunberg started the *Fridays for Future* movement by walking out of her school to strike in front of Sweden's parliament. After she garnered widespread recognition of her activism work, she was asked to speak at the World Economic Conference in Davos (Thunberg, 2019). In her speech she exclaimed, "Our house is on fire. I don't want you to be hopeful. I want you to panic. I want you to feel the fear I feel every day. And then I want you to act". Responding to Thunberg's call, on March 15, 2019, approximately 1.4 million young people walked out of their own schools in what might have been the largest eco-protest in history. They carried placards and wrote messages on their bodies such as, "Listen to Our Warning", "Climate is Changing Why Can't We?" "One Earth, One Chance", "Denial is Not a Policy", and "There is No Planet B" (Barclay & Amaria, 2019). If we bear witness to their anguish, Thunberg's pleas and the chants of the other students who filled the streets, it demands that we confront the social, political and economic structures fuelling the ecological emergency.

In Chapter 7, I sketch seven trajectories of ecosophical educational research, (EER) enmeshed and enlivened by the affective, cultural and ecological registers in my work. I hope the concepts and provocations shared, will connect with the passion, wisdom and creativity of other educational researchers to split open other trajectories. It is an attempt to find and connect with Others who believe educational research is not limited to making eco-attunement possible, but can be itself, the emergence of eco-attunement. The last section of the chapter identifies nine broad areas for educational researchers to consider if they are interested in pursuing intersubjective inquiries into the affective dimensions of curriculum and pedagogy amid environmental degradation.

In neoliberal frames, learning is associated with economic success and the acquisition of material goods. It dissuades students and teachers from critiquing the same economic system which is responsible for ravaging the planet. As the climate crisis intensifies, educational leaders will need to work through their own ambivalence and reconnect with the moral purpose of leading. In Chapter 11, I explain how ecosophical educational leadership research (EELR), can play a significant role in developing images of practice that focus on ethical leadership encounters inclusive of our more-than-human-relatives. I build on the seven trajectories of EER discussed in Chapter 10, and identify some areas of EELR grounded in radical hope and problematizing the asymmetrical communication processes that reify entrepreneurial frames of leadership. Especially at this time in our history, the beauty, complexity and unbidden mysteries of education should not be reduced to skill building, rubrics and checking off predetermined outcomes. Ecosophical education resists such a constrained visuality.

Informed by Relational Psychoanalysis

An important component of this work is to build a relational approach (Benjamin, 1998; Benjamin, 2018; Mitchell, 2000; Orange, 2011; Orange, 2017; Ringstrom, 2010) in *the field* of educational research, one that centres mutual recognition. The intersubjectivity of educational research, exists within a field that includes the total number of influences and the outcomes of those influences (Stern, 2015). Enmeshed in relations, we come to know the field through our senses. In other words, there is no outside. That said, I pay particular attention to the affective and aesthetic dimensions of education in the Anthropocene throughout the text. My intent is to open a space for monstrous feelings and to connect readers to the affective charges that move between subjects and objects within the field of relations. I define a monstrous creation as a creative work, creative workings or spectacles that bring about the uncanny (Freud, 1919). These sensuous and contradictory interruptions of and by the strange, produce a haunting vitality that disrupts unhealthy patterns in relational matrices.

To make a case for a turn to relational and psychoanalytically informed educational research, I discuss the evolution of intersubjective psychoanalytic theory by introducing the work of Sándor Ferenczi in Chapter 6. Building on Ferenczi's seminal work on mutual analysis, I explore the phenomena of transference and countertransference and how they impede mutual recognition in classrooms. I go on to expand Winnicott's concept of the holding environment by including social media's digital habitat and the more-than-human-world. As I broaden the field of relations, I identify some of the dialogical and material barriers exacerbating the political conditions that make a wide-scale response to ecological collapse exceedingly more difficult.

The intersubjective question posed in Chapter 7, is *how can I become my Other's keeper?* I query how the climate crisis intensifies mechanisms the privileged (gender, race, class, sexual orientation) use, to collude in the maintenance of systems of oppression. I draw on Levinas's work to make the case it is a moral obscenity to ignore the disproportionate impact climate change is having on minoritized people and I use a concept of the *ouroboros-subjectivity,* to describe the problematic ways in which humans determine their relative importance to Others.

The World Pressing In

Climate change and COVID-19 are rooted in the ecological exploitation of capitalism. To draw some lessons about what the pandemic revealed about our ability to address the climate crisis, in Chapter 8, I share some of the findings from a study called, *Educational Leadership during the COVID-19 Outbreak.* The aims of the research project were to understand the complex challenges educational leaders were facing in the first phase of the pandemic; to illuminate the affective dimensions of leading in precarious times; and to study what leading during the pandemic could teach us about educational leadership, and the field of education more broadly amid the climate crisis.

The pandemic produced differentiated risks and exposures across school communities. Managerialist and technocratic leadership fell flat in attempts to mitigate the challenges experienced by minoritized students. Through mapping their stories affectively, rather than managerially, the leaders I spoke to became less fearful of living out their ethical and moral understandings of the pandemic together. When they exposed their vulnerability to Others, they found the space was riddled with redemptive possibilities and the perils of misrecognition. Nonetheless, they concluded, we all have a better chance of emerging from a crisis psychologically intact, when we become more accountable for our affective attachments and our obligations to the Other.

Posing questions from different registers of life (e.g., asking ourselves what insects have to teach humans about living in balance with other Beings) can disrupt systems of entrenched power and shift us into new assemblages. To that end, this work places us on rough ground. In the first place, education and psychoanalysis (Freud, 1913; Gentile, 2018) have been historically entrenched in anthropocentric assumptions. Psychoanalysis is understood as *the talking cure*. The analyst privileges the spoken word of humans, and since the time of Freud, has endeavoured to exorcise animalistic instincts (picture Freud's Wolf man) from the consulting room. White-male-settlers are at the centre of western education's project of dominance. The colonial and patriarchal architecture treats animals and plants, not as ends in themselves, but as resources that make the products people are socialized to consume.

My work has always maintained a strong focus on the relational and affective dimensions of teaching, leading and learning. A central component of my first book (Farrell, 2020) was a three-act play called, *Last Call for Sincere Liars*. The play centres on the devolution of a school principal named Martin and the affective register of his life. In the play, the field is represented by the pub in which Martin explores his dis/connection to his bartender-analyst, Karen. The pub is overtaken by an unnamed destructive force called, *The Grey* in the last act. A few friends and colleagues I called upon to provide me with critical feedback on the third act said it was, "excessively gloomy", "a bit too dramatic", and "too sad because it all goes away for Martin in the end. The phantoms eat up his world". In response to their feedback, I rewrote the third act several times, each time trying to inject it with a few happy affects, failing in each attempt. It is true, the ending is pretty gloomy. Martin is not gifted with an epiphany to keep The Grey at bay. The characters are wild affects, oscillating inside their human shells as the vestiges of unprocessed trauma. What is more apparent to me now, is writing the play, invited my monstrous worries about the end of the world to come closer. It was a way to confront my fear of the "phantoms" that churn the natural world into capital in service of the economy that will "make it all go away in the end".

That said, this has been a challenging project. Over the past few years, there were days when I was unable to write, frozen by the most recent story of environmental degradation popping up as a notification on my phone. On other days, I wrote in a mad panic, driven by a desire to pen something of consequence before time is up. And then, there were the awful sit-in-a-ball-in-the-corner-of-my-office-days, when the melancholy set in during times of pandemic reclusion. Thankfully, walking among the birch trees, taking photographs of black bears eating raspberries in Riding Mountain National Park and capturing images of terns feeding each other silver fish, kept me open to the unbidden. In the company of my more-than-human relatives, I continue to be restored. T. S. Eliot (1944), once said, "Humankind cannot bear much reality" but I remain radically hopeful of what education can do to help all of us bear the difficult knowledge of the climate crisis while finding ourselves as our Other's keepers.

References

Abraham, N., & Torok, M. (1971). The topography of reality: Sketching a metapsychology of secrets. In Abraham, N., & Torok, M. (1994). *The shell and the kernel: Renewals of psychoanalysis* (pp. 157–164) (Vol. 1). (N. T. Rand, Trans.). University of Chicago.

Abraham, N., & Torok, M. (1972). Mourning or melancholia: Introjection versus incorporation. In Abraham, N., & Torok, M. (1994). *The shell and the kernel: Renewals of psychoanalysis.* (pp. 125–137) (Vol. 1) (N. T. Rand, Trans.). University of Chicago.

Antonioli, M. (2018). What is ecosophy? *European Journal of Creative Practices in Cities and Landscapes.* 1–8. https://creativecommons.org/licenses/by/4.0/

Barclay, E., & Amaria, K. (2019, March 17). Photos: Kids in 123 countries went on strike to protect the climate. Vox. https://www.vox.com/energy-and-environment/2019/3/15/18267156/youth-climate-strike-march-15-photos

Benjamin, J. (1998). *Shadow of the other: Intersubjectivity and gender in psychoanalysis.* Routledge.

Benjamin, J. (2018). *Beyond doer and done to: Recognition theory, intersubjectivity and the third.* Routledge.

Bhalla, J. (2021, February 24). What's your "fair share" of carbon emissions? You're probably blowing way past it. Vox. https://www.vox.com/22291568/climate-change-carbon-footprint-greta-thunberg-un-emissions-gap-report

Boin, A., 't Hart, P., Stern, E., & Sundelius, B. (2017). *The politics of crisis management: Public leadership under pressure* (2nd ed.). Cambridge University Press.

Citton, Y. (2014). *Pour une écologie de l'attention*, Seuil, Nanterre.

Cunsolo, A., Harper, S. L., Minor, K., Hayes, K., Williams, K. G., & Howard, C. (2020). Ecological grief and anxiety: The start of a healthy response to climate change? *Planetary Health, 4,* e261–e263. doi: https://doi.org/10.1016/S2542-5196(20)30144-3

Dickinson, E. (1896). Time and Eternity, Poem 29: Ghosts. *The Poems of Emily Dickinson: Series Two* (Lit2Go Edition). Retrieved May 13, 2021, from https://etc.usf.edu/lit2go/115/the-poems-of-emily-dickinson-series-two/4546/time-and-eternity-poem-29-ghosts/

Eliot, T. S. (1944). Burnt Norton. In Four Quartets. Faber.

Farrell, A. J. (2020). *Exploring the affective dimensions of educational leadership: Psychoanalytic and arts-based methods.* Routledge.

Freud, S. (1913). Totem and Taboo. Norton (1989).

Freud, S. (1919). *The uncanny.* [Kindle Edition]. Retrieved from amazon.ca

Gentile, K. (2018). Animals as the symptom of psychoanalysis or, the potential for interspecies co-emergence in psychoanalysis. *Studies in Gender and Sexuality, 19*(1), 7–13. https://doi.org/10.1080/15240657.2018.1419687

Greenhalgh-Spencer, H. (2014). Guattari's ecosophy and implications for pedagogy. *Journal of Philosophy of Education, 48*(2), 323–338. https://doi.org/10.1111/1467-9752.12060

Guattari, F. (2008). *The three ecologies*. (I. Pindar, & P. Sutton, Trans.). Continuum.

Hardt, M., & Negri, A. (2017). *Assembly*. Oxford University Press.

Heinz, L. (2017). *Extinction: A photographic exploration by Marc Schlossman*. https://www.extinction.photo/about-the-extinction-project/

Hickman, C. (2019). Children and climate change: Exploring children's feelings about climate change using free association narrative interview methodology. In P. Hoggett (Ed.), *Climate psychology: On indifference and disaster*. (pp. 41–59). Palgrave Macmillan.

Hickman, C. (2020). We need to (find a way to) talk about … eco-anxiety. *Journal of Social Work Practice, 34*(4), 411–424. https://doi.org/10.1080/02650533.2020.1844166

Manitoba Education. (2006). *Grade 11 chemistry: A foundation for implementation*. In Manitoba Education CaY, editor. Winnipeg, Manitoba. p. 159.

Mitchell, S. A. (2000). *Relationality: From attachment to intersubjectivity*. The Analytic Press.

Naess, A. (2008). *The ecology of wisdom: Writings by Arne Naess*. A. Drengson & B. Devall (Eds.). Counterpoint.

Orange, D. (2011). *The suffering stranger: Hermeneutics for everyday clinical practice*. Routledge.

Orange, D. (2017). *Climate crisis, psychoanalysis, and radical ethics*. Routledge.

Ringstrom, P. A. (2010). Meeting Mitchell's challenge: A comparison of relational psychoanalysis and intersubjective systems theory. *Psychoanalytic Dialogues, 20*, 196–218. doi: 10.1080/10481881003716289

Ruti, M. (2018). *Distillations: Theory, ethics, affect*. Bloomsbury.

Stern, D. B. (2015). *Relational freedom: Emergent properties of the interpersonal field*. Routledge.

Thunberg, G. (2019, January, 25). 'Our house is on fire': Greta Thunberg, 16 urges leaders to act on climate. The Guardian. https://amp.theguardian.com/environment/2019/jan/25/our-house-is-on-fire-greta-thunberg16-urges-leaders-to-act-on-climate

United Nations Environment Programme. (2020). Emissions gap report 2020. https://www.unep.org/emissions-gap-report-2020

Wynes, S., & Nicholas, K. A. (2017). The climate mitigation gap: Education and government recommendations miss the most effective individual actions. *Environmental Research Letters, 12* (7) https://doi.org/10.1088/1748-9326/aa7541

1 Renovating the Schoolhouse When the World is on Fire

Fire is often used as an image to represent the existential threat global warming presents. Klein (2019) uses the image of a raging fire to describe young people's apocalyptic fears about the degradation of the Earth's ecosystems. In reference to the thousands of young people who participated in the first *Global School Strike for Climate*, Klein states, "Suddenly, children around the world were taking their cues from Greta [Thunberg], the girl who takes social cues from no one, and were organizing student strikes of their own. At their marches, many held up placards quoting some of her most piercing words: "I WANT YOU TO PANIC, OUR HOUSE IS ON FIRE!" (p. 16). Whenever images of flames or smoke are evoked to refer to the climate crisis, a repetitive trail of thoughts grips me. I wonder what kind of people light the curtains on fire and then stand on hope and faith the children will get out of the house in time. Dread follows. There is *no outside* of our Earth home. A vice grip tightens around my chest, like it did when I was a classroom teacher responsible for hurrying students down the hall, lining them up outside and then counting their heads as the fire alarm rang in our ears. These bodily sensations often cascade into a reoccurring dream of a group of teachers pulling a fire alarm and then running outside. In my dream, thousands of singed blue attendance cards blow along the playground's chain link fence. I turn to count my students' heads, but they are missing. My cheeks flush. I glance over to see an old woman flick a lit cigarette on the ground. She tosses me her cigarette package and declares, "Don't you worry. They'll get out in time".

In Canada, every education facility is expected to have a fire safety plan. The bulk of the plans emphasize emergency planning, training staff and students on what to do in case of a fire and keeping the building fire safe (Manitoba School Boards Association, 2018, p. 1). Fire departments and inspectors work with school administrators to plan safety routes, designate staff roles and to ensure regular fire drill practices occur throughout the school year. There is widespread agreement among the public that schools present distinctive fire and safety risks. Communities expect schools to receive outside expertise, adequate information, clear

DOI: 10.4324/9781003024873-2

instructions and financial support in order to create safety plans, mark escape routes and to buy and maintain the requisite equipment such as fire alarms and exit signs. It would sound ludicrous to hear political pundits, policymakers or leaders quibble over the resources, time or training needed to develop a well thought out fire safety plan.

When it comes to fire safety, even in the Wild West of late-stage capitalism, our visuality expands to see more of the wider web of relations supporting the lifeworld of the school. Water, wind, fire hydrants, fire fighters, inspectors, emergency response systems and 911 operators are integral elements of a *school's life-web* when writing emergency response plans or fighting a fire in the chemistry lab. When it comes to the climate emergency, we must also expand our reductive understandings of what constitutes the school community. The *school* is so much more than its brick and mortar façade. We must zoom out from individual buildings and take the creatures, land, water, plants, cultural practices and social systems into account. An expansion of what constitutes *the school* has ecosophical implications as well as lifesaving potential.

If we were to try and mitigate the damage to school communities caused by wildfires, we would consider the climate, including temperature, precipitation and atmospheric moisture, as critical elements of fire activity (Wang et al., 2015) and incorporate this knowledge into fire safety plans, pedagogy and curricula. Climate determines the amount and type of vegetation (or fire fuel) in a given location. As global warming increases, there are longer fire seasons, more lightening and drier fuels. In spite of this knowledge, in Canada, a country heating two times faster than most other places on Earth (Zhang et al., 2019), a country that is literally catching fire more often because of climate change, there is nothing but quibbling about how best to respond to the threats posed by the climate crisis. To be sure, the wealthy elite in the west already smell smoke. They just refuse to pull the fire alarm because they need both hands to cling to their pearls and gold. Alternatively, there are many places in the world where the smoke is thick and the fire is raging. And tragically, as more and more of our Earth house literally catches fire, more of the Earth's precious resources are stolen from the poor and reallocated to the rich.

In this chapter, I speak to the real and metaphorical fires in our current context, the ones making it difficult to reorient education and research towards ecosophical aims. I begin with the boil water advisories in rural and remote First Nations communities in Canada and then discuss the Flint, Michigan, water crisis to describe an erosion of the spirit of Mni wiconi by Mad Max logic. Then, I discuss how the brutality of late-stage capitalism leads to educational policy and school renewal efforts that are antithetical to compassionate witnessing and an attunement to the more-than-human-world. I go on to consider how the traditional media and social media shape climate change discourse for better and for worse.

Mad Max versus Mni Wiconi

On December 6, 2020, Bloomberg reported California's water futures were officially released onto the Wall Street markets. The Royal Bank of Canada's managing director and analyst, Deane Dray, told Bloomberg news, "Climate change, droughts, population growth, and pollution are likely to make water scarcity issues a hot topic for years to come. We are definitely going to watch how this new water futures contract develops" (Martin, 2020). As Martin aptly describes in his coverage of Bloomberg's chilling report, "We've officially reached a new phase of the Mad Maxification of America" (Martin, 2020, para. 1). The Mad-Maxification of water in Canada, flows in the toxic water supplied to many Indigenous communities. The Canadian government disclosed in January 2018, 91 First Nations were under long-term water advisories (Amnesty International, n.d.). In addition to the federal government's admissions, several human rights groups over the last three decades have raised the alarm because the water in many Indigenous communities is contaminated or at risk due to defective or inoperative treatment systems (Human Rights Watch, 2016).

Responding to public pressure from Indigenous leaders, activists and NGOs, Prime Minister Justin Trudeau made a pledge in 2017 on World Water Day to end the need for water advisories in Canada (Aiello, 2017). Indigenous leaders from across the country were not surprised, however, to hear the Indigenous Services Minister (ISM) confirm in December of 2020, the federal government would once again fail to keep its promise (Russell, 2020), citing the COVID-19 pandemic as the most recent barrier. The announcement from the ISM was particularly abhorrent because Health Canada initially declared persistent handwashing with soap and clean water was key to preventing the transmission of COVID-19. Unsafe drinking water in many rural and remote Indigenous communities is evidence that water in North America is viewed as a commodity that can be acquired and hoarded through colonial, capitalist and white supremacist structures of oppression.

In 2011, the state of Michigan in the United States took over the city of Flint's finances after an audit projected a 25-million-dollar deficit (CNN, 2020). To address the shortfall in the water budget, the city switched its drinking water supply from the Detroit system to the Flint River. The water was not tested nor treated properly, and soon after the switch was made, residents began to complain the water was foul-smelling and discoloured (Denchak, 2018). The city's politicians, contractors and water company officials who machinated the change, knew there could be dangerous levels of lead pouring from the taps in people's homes (Holden et al., 2019). A coordinated campaign of denial ensued to mute the cries of parents and physicians who were alarmed at the affects the lead poisoning was having on the bodies of Flint children, including hearing problems, cognitive disabilities and significant behaviour changes.

Even before the Flint water crisis made international news, for more than 100 years, the river running through the city was used as a disposal site for lumber mills, meat packing plants and raw sewage from the city's waste treatment plant. The Michigan Civil Rights Commission published a report citing systemic racism as the root cause of the Flint water crisis (Karoub, 2017). Flint, which is a predominantly Black city, was forced to use the poisonous water from the Flint River for 18 months without it being treated to prevent pipe corrosion. As a result, lead leached from old pipes and into water glasses and cooking pots. The Flint water crisis is yet another disturbing example of how the lives of racialized and minoritized people are severely impacted by eco-racist policies.

"Mni wiconi" means "water is life", in the language of the Lakota. Water flows through the interconnections among all living beings on Earth. As they watch the climate crisis intensify, the moneyed are beginning to covet fresh water. Wall Street's hijacking of California's water futures means traders have predicted with confidence that there will be widespread reductions in the quality and quantity of drinkable water in California. The Mad Max logic of critical resource shortages, hyper individualism and the colonization of the commons by financial marauders is breeding social acceptance of an inevitable devolution into barbarity and the shredding of the social contract. The erosion of Mni wiconi among the masses by Mad Max logic is one of the most vulgar aesthetics of late-stage capitalism.

The Brutality of Late-Stage Capitalism

Since the 1990s, the number of severe climate events has doubled. During the same period, researchers marked increases in post-traumatic stress disorder, depression, anxiety, drug use and domestic abuse among adults (Kurth, 2017; Van Susteren & Colino, 2020). The executive summary of the Psychological Effects of Global Warming on the United States forewarns that the climate crisis "...in the coming years will foster public trauma, depression, violence, alienation, substance abuse, suicide, psychotic episodes, post-traumatic stress disorders and many other mental health conditions" (Coyle & von Susteren, 2012, p. i). Researchers have begun to investigate the links among climate change, rising temperatures and the suicides of Indian farmers over the last three decades. In the 35 districts of Maharashtra, the number of farmers' suicides doubled to 11,995 in the last four years (Purohit, 2019). In another study, Carleton (2017) found changes in temperature were related to suicide rates in India. For temperatures above 20 degrees Celsius, a one-degree Celsius increase in a single day's temperature was linked to on average 70 suicides. She determined this effect occurred only during India's agricultural growing season when heat also lowered crop yield (p. 8746). The dire situation of farmers in India is reflective of the deep connection between climate precarity and the fragility of psychological health in times of prolonged crisis.

We need a deeper understanding of the trauma responses precipitated by the climate crisis in the field of education. Much of the trauma informed work happening in schools operates within a medicalized model which obfuscates the structural oppressions embedded in neoliberal politics, polices and governance structures. A significant impasse to the intersubjective study of trauma in education is the dominance of cognitive behavioural therapeutic (CBT) theories and techniques (Taubman, 2012). In broad strokes, the aims of CBT are to assist individuals to alter cognitive distortions, increase capacity for emotional regulation, change negative coping behaviours and to develop personal coping strategies (Hoogsteder et al., 2015). Three common CBT techniques used in schools are self-talk, the articulation of SMART (Specific, Measurable, Attainable, Relevant, Time-bound) goals and cognitive restructuring. Within the context of the traumatic dimensions of the climate emergency, the dominant discourse of CBT falls short in terms of being able to generate shared meaning of the crisis. Instead of addressing the social, structural and cultural causes of anxiety and depression, diagnoses and treatments are directed at the individual. CBT does not actively trouble the harsh reality of the neoliberal project and its deleterious mental health effects on its subjects. In fact, due to its orientation towards modifying the behaviour of individuals, CBT obfuscates the culturally contingent terrain of *telluric emotions, emotions that are of the Earth, shared and mutually influencing.*

Neoliberalism and the Culture of Austerity

The public in public education is disappearing in cultures of manufactured austerity by cruel budget cuts, couched in the language of increased efficiency and greater accountability. Each time the public education system is characterized by right-wing politicians as bloated, ineffectual and protected by self-interested teacher unions, the public's confidence is shaken and more people become susceptible to the neoliberal fantasy that whatever needs to be done, the private sector can do it better. Starve a system. Make it weak. Vilify the people in the system for doing a poor job. Offer an alternative from the private sector. Repeat. Dufresne (2019) describes the wider phenomenon as *boomernomics*, which is:

> "...the shift from production and need to consumption and desire; the high value assigned to audit culture and metrics; the quantification of job performance; the belief that efficiency and continuous improvement are good in themselves; the reverence given to strategic

planning, mission statements, outcome and managerial expertise, essentially as legitimized by the MBA degree; the dogma of tax reductions, hatred of "big government," and the de facto institutionalization of trickle-down economics... and so on *ad nauseam*.

(p. 52)

Many right-wing political leaders who pursue draconian changes to public school systems are quite adept at activating and manipulating the affects-of-late-stage-capitalism. For example, the *siren song of grievance* sung by right-wing provocateurs is intended to promote aggressiveness and competition, magnify divisions among groups of people, separate humans from nature and suppress critical reflection. The success of right-wing politicians in designing emotionally charged narratives to work against the public interest are due in part to the support they receive from corporate lobby groups and think tanks. The provincial government in Manitoba was recently accused of plagiarizing legislation from the American Legislative Exchange Council (ALEC). ALEC "works behind the scenes to provide fill-in-the-blanks legislation to Republican legislators to promote their corporate right-wing agenda: making government as small and taxes as low as possible, bringing in anti-union right-to-work acts, voter suppression laws and even stand-your-ground NRA progun legislation" (Manitoba Organization of Faculty Associations, 2021). To manufacture public consent for reforms birthed in right-wing think tanks, broad-scale reviews and audits are frequently used to justify predetermined changes to the financing and governance of school systems (Wilkins et al., 2021).

In January 2019, the Manitoba government announced that a nine-member commission would facilitate a comprehensive review of the K-12 education system. Five days before the commission released their report, the Minister of Education, Cliff Cullen, held what was described as "perplexing press conference" (Bernhardt & Frew, 2021). When questioned by reporters, he refused to provide any details regarding the major changes the government planned to make. Instead, the minister made several inflammatory remarks, characterizing the school system as a failure. The press conference was used as a mechanism to deliver a preemptive psychological jolt to the public and to weave a fantasy in which the incongruities and inconsistencies of the education review could be overcome. The fact that Manitoba has the second highest child poverty rate in Canada, which is the most significant factor in determining student success, was notably missing from the minister's remarks.

In the October 2020 Throne Speech, the government outlined its "Better Education Strategy Today (BEST)" plan (Manitoba, 2020). BEST promises the government "...will improve outcomes and achievement for all students across the province", "...equip teachers and administrators with the tools they need to be successful" and "...build consistency and coherence across the province, with stronger accountability structures" (Manitoba, 2020). A colleague of mine from the University of Manitoba analyzed the priorities of the BEST report reflected in word usage and frequency (Janzen, 2021). She found the word *performance* appeared 31 times, *achievement* is referenced 22 times and *employability* appears 12 times. When centralizing a data-driven approach to close achievement gaps, labour and market capital become the government's proxy for equity. The word democracy never appears in BEST and much of the discourse in the report severs public education from the aims of social justice and critically engaged citizenship.

Coverage of the Climate Crisis

Twenty years ago, climate was rarely front-page news. Extreme weather events and the insect apocalypse were the stuff of cli-fi movies and dystopian fiction. In the last five years, reporting on global warming has significantly increased. As a corollary, research about the type of coverage and its relationship to public perception has bloomed. Researchers are monitoring the number of reports published and how particular media frames influence public perceptions about the drivers and relative urgency of the climate crisis. The research shows countries differ in several important ways with regards to what and how they cover climate change. In the United States, there has been an overall increase in media coverage. In Canada, coverage of the anthropogenic causes of climate change is relatively infrequent (Tschötschel et al., 2020). When climate change is covered in Canada, media reports typically focus on the physical impacts of climate volatility. For example, in August 2020, several media outlets reported the last fully intact ice shelf in the Canadian Arctic collapsed. It lost 40% of its area, approximately 80 square kilometres in two days at the end of July (Warburton, 2020). Over the last three decades, the Canadian arctic has been warming at twice the global rate, and as more glaciers disappear, more of the bedrock in the arctic is exposed to heat. The heat from the newly exposed bedrock is further accelerating the melting process.

In 2020, the Pew Research Center reported that for the first time in its surveys dating back two decades, "nearly as many Americans say protecting the environment should be a top policy priority (64%) as they say this about strengthening the economy (67%)". Researchers posit that as concern for the economy receded in people's minds, concern for the environment increased (Svoboda, 2020). Relative to the United States, Canadian media reports have focused more on the health impacts related to climate volatility and less on the economic effects. A quantitative spatiotemporal analysis of Canadian newspaper coverage of climate change impacts on health between 2005 and 2015 found that almost all the 145 articles described negative impacts on health (King et al. 2019) emphasizing chronic non-infectious and infectious diseases (p. 581).

The media's emphasis on controversy versus consensus also influences public attitudes and assumptions about climate change (Tschötschel et al., 2020). For example, media reports emphasizing controversy among scientists about the anthropogenic causes of global warming are linked to lower levels of policy engagement and public support (Lorenzoni et al., 2007; van der Linden et al., 2015). Another media study examined the disparity in post-disaster reporting of two events: Hurricane Harvey and the South Asian floods in the Global South and Global North in UK newspapers. Reporting bias was found in the number of pieces published, the length of articles and the duration of coverage (Porter & Evans, 2020, p. 355). More significantly, the studies found media reports from the UK about the Global South reinforced an "us versus them" mentality, potentially suppressing solidarity and momentum for collective action.

To generate a small sample of climate change coverage, I read and then gathered headlines from the first two weeks in January 2021 from three different news organizations. In Appendix A, you will find a text block of the headlines or standfirsts from each news article posted in the climate change section of CBC's, Aljazeera's and The Guardian's websites. I selected either the headline or the standfirst from each of the articles based on which of the two best reflected the content and context of each article. As you scan the text block, I encourage you to take notice of the words, phrases and images that give you pause, quicken your pulse, flood your eyes, flush your cheeks or make you want to click. You might hold your breath while you read into the headline "the sea is rising" or feel chills ripple up your spine as you begin to process the words "Top scientists warn of 'ghastly future of mass extinction'". Maybe a tingle of the uncanny wraps around you because the headline "Why does Australia act as if it can ignore the climate crisis, and how long can it keep to this seemingly suicidal posture?" makes you question whether or not countries can be suicidal. The 61 headlines and standfirsts offer a snapshot in time of the mediated climate information assemblages that circulated online in the beginning of January 2021.

In terms of topic frequency, you will notice in Appendix A that increases in global temperatures and economic impacts were the focus of 11 articles each, accounting for 34% of the articles. Only 8% of the articles centred on the impact of climate change on other Beings and only 10% of the articles discussed the degradation of ocean ecosystems. In this tiny data set, anthropocentrism, rising temperatures and the economy are amplified and the impacts on the-more-than-human world and the precarity of ocean ecosystems is given less attention. Of course, parsing headlines by the numbers will not take us where we need go even if we grappled with a sample size that was infinitely larger. We need to read into the spaces between the words in the headlines, climb inside the particularities. For instance, the headline "Climate scientist says that another top 10 year is a 'no shit, Sherlock' moment as temperatures across the country were 1.12C above average" invites us into the psychoanalytic realm. Are we delusional? Why are so many humans so calm in the face of such calamitous projections? The "no shit Sherlock" headline is a provocation to wonder about the psychological defences at work in the collective psyche.

The Antisocial Digital Context

Three years ago, I facilitated a small study to explore how teacher candidates curate their identities on various social media platforms. I wanted to learn more about relational encounters when they are mediated through social media platforms. I interviewed eight teacher candidates. One of the key insights from the study was how the participants used the process of *digital echolocation* to make meaning of the disparity between what they perceived as their *real selves* and the multiple iterations of their identities across different social media platforms. When the distance between the real self and the social media avatars they curated became too great, loneliness and insecurity contributed to the corruption of digital echolocation.

One of the participants, Marta, engaged continuously in digital echolocation as she navigated the complexity of the online gaming world as a woman. She pronounced:

> ...And typically like the long-term strategy users, will just fuck with people. I try to figure out things like, so are they just interested in meeting people, or are they, are they trying to like tell their views to you, are they just like, are they trying to figure out what your strategy is? Um, so that, like that parts kind of fun, like it becomes the messaging and building relationships almost becomes an entire strategy game of its own.

Marta, along with other participants, described social media as a game within a game. On the one hand, the participants were ambivalent about the quality of their online relationships and on the other, they were hyper focused on gaining approval from particular followers or friends while trying to avoid being labelled as unprofessional by someone who would have some say in hiring or firing them. What was most compelling is that most of the teacher candidates seemed to have already internalized an administrative figure to police their behaviour and to defend against negative feelings when everyday cruelties were amplified online. The relational psychoanalytic lens I used to look at the interview data troubled the participants' descriptions of the existence of an intact, isolated, stable and core self. In fact, what the participants described were a multiplicity of identities simultaneously extended across and influenced by multiple fields of relations.

This small study triggered new curiosities about digital echolocation as it relates to how users engage with the topic of climate change and the suffering stranger on social media. Platforms like Facebook and Twitter are used extensively during and after a disaster. Some of the videos, photos or messages users exchange are used to locate survivors, warn people of new dangers or to organize emergency relief efforts (Alam et al., 2018). More recently, the proliferation of climate emergency images is driven by an appetite for disturbing images. Disaster porn is lucrative clickbait, but it separates the images of burnt trees, flooded basements, oil-soaked wings and bloodied faces in Red Cross tents from the larger systemic problems that precipitate the crises. Users digitally echolocate themselves into a stranger's world to observe and *try on suffering* from a distance. Voyeuristic and exploitative, the acritical consumption of these images churns pain into a commodity and worse, it desensitizes the user, who in turn seeks out more dramatic footage to satisfy the next craving. Disturbing images of the impacts of the climate crisis are no exception.

"People have long been drawn, purposely or otherwise, towards sites, attractions or events linked in one way or another with death, suffering, violence or disaster" (Stone & Sharpley, 2008, pg. 574). In September 2015, the Turkish photojournalist, Nilufer Demir took a photograph of the body of a three-year-old child who had drowned when his family fled the war in Syria. The photograph went viral. First, the image circulated as a media spectacle and then it was reconstituted as a variety of memes on social media (Durham, 2018). The acritical consumption and exchange of Demir's photo and other shocking images of death, destruction and suffering is an exercise in dark tourism (Lonergan, 2020). Twitter, Instagram, YouTube and Facebook jettison digital tourists into faraway places to experience the strange, horrific or the exotic from a safe and detached distance. The most painful moments of a suffering stranger's life mutate into an affectively charged visual feast. With each retweet, share and like, the most disturbing images recirculate with greater intensity through the digital circulation system.

Algorithms that link social media posts to purchasing histories and political affiliations, the same algorithms ensuring we are not troubled by opposing viewpoints have made the digital ecosystem a place to enhance one's epistemic vices at the expense of democratic praxis (D'Olimpio, 2021; Franch, 2020). And despite the proliferation of disturbing images and the deluge of antisocial commentary, digital citizenship is often constructed as the skills and competencies an individual student must acquire in order to safely and productively navigate online spaces (Buchholz et al., 2020). An intersubjective positioning of the complexity of our digital identities and the cultivation of a shared set of values is often absent in digital spaces of curriculum and pedagogy. For example, when schools and universities shuttered during the COVID-19 pandemic, classes moved online. Many educators used the Zoom platform to teach via video conference. Soon after, reports of Zoombombing began to circulate in the media, a violation "…in which intruders hijack video calls and post hate speech and offensive images such as pornography" (Bond, 2020, para. 1). When incidences of Zoombombing happened to my colleagues, their universities or school divisions approached the intrusions as security breaches and technical problems. Accounts and video meetings were fortified with passwords and protocols were shared about what educators and students ought to do when classes were Zoombombed.

If one of the aims of education is to learn to live relationally, discussions of digital citizenship in educative encounters must include the dynamic interplay of identity development across screens and platforms, critical analyses of injustice, civic and civil discourse, accessing and sharing dis/information and the relationship between guttural responses and the curation and datafication of our lives. Guattari's (2008) plurality of ecologies suggests the environment is not the fishbowl that surrounds the goldfish. It is a dynamic network of relations we define and define us in perpetuity. The definitive lines we arbitrarily draw to separate humans from their digital worlds do not hold. Not only do the illusory demarcations make it more challenging to notice the dynamism of life, it circumvents the enactment of multi-vocal dialogical spaces needed to move towards transformation. It is the organic fusions and dissolutions of a multiplicity of connections among Beings that create the moments of radical psychosocial experimentation needed to respond to the most pressing issues of our time. If we breach the illusory lines separating the digital world from the real in moral terms, new trajectories open up. We can find ways to live more relationally within our digital cells. In online educative encounters, educators and students could focus less on skill-building and focus more on ethical and moral engagements in political and social activities. Digital citizenship, then, must be reframed as participatory student engagement and as a pathway to interrupt the objectification, exoticization and marginalization of those impacted by systemic oppression and eco-injustice.

An Urgent Call to Our Better Angels

Anthropocenity is a conscious awareness of the fact that we are living in a time of black swan events. Humans are less confident that they can rely on the natural world to provide them with what they need to live. The anxiety is palpable on social media, around the dinner table and in the lecture hall. It emanates from a growing awareness of how the political, economic and cultural mechanisms of late-stage capitalism makes the lives of most Beings on Earth expendable. One of the most well-known articulations of the cultural embeddedness of such a corrupt system comes from Pope Francis in *Laudato Si'* (2015). He calls the climate emergency a "spiritual crisis in modernity" that stems from:

> the way that humanity has taken up technology and its development according to an undifferentiated and one-dimensional paradigm. This paradigm exalts the concept of a subject who, using logical and rational procedures, progressively approaches and gains control over an external object. This subject makes every effort to establish the scientific and experimental method, which in itself is already a technique of possession, mastery and transformation.
>
> (106)

Human techniques of possession, mastery and transformation have led to the desertification of land and the displacement of thousands of people. The displacement of millions of people is causing massive upheavals in social and cultural systems (Hoggett, 2019). More than half of the original area of forest and grassland terrestrial ecosystems has been converted to farmland, making it harder to rewild much of the planet. Inuit in Labrador, Canada, are experiencing existential distress as they witness the ice that is deeply connected to their identity disappear (Wray, 2019). Animals cannot adapt fast enough in ecosystems altered by rising temperatures and the sixth mass extinction is picking up speed (Kolbert, 2014). While the impacts of the climate crisis are often decontextualized from human choices and actions, the destructive consequences of treating relations with more-than-human-world in terms of resource extraction are becoming intolerable.

Suzuki (Cape Farewell, n.d.) puts it succinctly, "My postwar generation and the boomers who followed – we've lived like kings and queens, and we partied like there is no tomorrow, never worrying about the kind of world we were leaving for our children. Well, the party's over. Human beings and the natural world are on a collision course". The Intergovernmental Panel on Climate Change (IPCC) is the United Nations' body evaluating the science related to climate change. Its latest report echoes Suzuki's dire prediction. We are careening down the "collision course" faster than predicted. In 2018, the IPCC produced a report on the impacts of global warming of 1.5 degrees Celsius above pre-industrial levels and related global greenhouse emission pathways (IPCC, 2018, pg. 4). The report states that human activities are estimated to have caused approximately 1.0 degree Celsius of global warming above pre-industrial levels and that global warming is likely to reach 1.5 degrees Celsius between 2030 and 2052, if it continues to increase at the current rate (IPCC, 2018, pg.4). As a result, climate-related risks to health, livelihoods, food security, water supply, human security and economic growth are projected to increase with global warming of 1.5 degrees Celsius and increase further with 2 degrees Celsius (p. 9). The headline statements from the IPCC's Sixth Assessment Report (2021) that were released as this book was heading off to print are even more alarming. The United Nations Chief described it as "a code red for humanity" (McGrath, 2021).

Climate change is a cascading crisis impacting every aspect of our lives. The time for incremental change is over. We must be courageous and make substantial changes to our lives if we have any hope of turning things around in one generation. We can do it if shared purpose and the dignity of all Beings is forefront. At a time when the effectiveness of teaching is constructed mainly in terms of its capacity to feed the ends of capitalism, a turn to ecosophical education may seem irrational to the technocrats. Environmental activists might view the attunement the emotional dimensions of climate education as a luxury we cannot afford, but I believe a relational orientation in the field will create conditions in which education can play a fundamental role in the fight for a livable planet.

Even though the situation is dire, we have good reason to remain hopeful. Indigenous voices are coalescing across the globe to challenge colonial discourses. Teachings about the intricacies of the web of life and our moral, spiritual and ethical obligations to future generations are rooting in new places. McKibben, 2019; McKibben (2006), founder of the environmental organization 350.org, offers some additional reasons to remain hopeful. He reminds us that we already know what to do to mitigate the climate crisis, that is, we must keep fossil fuels in the ground and transition quickly to renewable energy. He goes on to note that renewable energy is getting cheaper every day and is heartened that the worldwide climate change movement is getting stronger. I, too, remain radically hopeful. This time of great adversity could cause a life-enhancing reimagination of education and educational research in which the suffering of Others is brought into sharper relief. Educators and educational researchers could play an important role in critiquing the current context and in orienting the field towards ecosophical ends.

References

Aiello, R. (2017). *Can PM Trudeau keep drinkable water promise to First Nations?* CTV News. https://www.ctvnews.ca/politics/can-pm-trudeau-keep-drinkable-water-promise-to-first-nations-1.3736954

Alam, F., Ofli, F., & Imran, M. (2018). Processing social media images by combining human and machine computing during crises. *International Journal of Human-Computer Interaction, 34*(4), 311–327. https://doi.org/10.1080/10447318.2018.1427831

Amnesty International (n.d.). The right to water. *Amnesty International.* https://www.amnesty.ca/our-work/issues/indigenous-peoples/indigenous-peoples-in-canada/the-right-to-water

Bernhardt, D., & Frew, N. (2021, March 12). Manitoba educators frustrated they won't see long awaited K-12 education system review until Monday. CBC News. https://www.cbc.ca/news/canada/manitoba/cliff-cullen-education-manitoba-1.5947230

Bond, S. (2020, April 3). A must for millions, Zoom has a darkside—and an FBI warning. NPR. https://www.npr.org/2020/04/03/82612 9520/a-must-for-millionszoom-has-a-dark-side-and-an-fbi-warning

Buchholz, B. A., Dehart, J., & Moorman, G. (2020). Digital citizenship during a global pandemic: Moving beyond digital literacy. *Journal of Adolescent & Adult Literacy, 64*(1), 11–17.

Cape Farewell. (n.d.). *The trial of David Suzuki.* https://capefarewell.com/explore/763-the-trial-of-david-suzuki.html

Carleton, T. A. (2017). Crop damaging temperatures increase suicide rates in India. *Proceedings of the National Academy of Sciences of the United States of America, 114*(33), 8746–8751. https://www.pnas.org/content/pnas/114/33/8746.full.pdf

Chipman, K. (2020, December, 6). California water futures begin trading amid fears of scarcity. *Bloomberg Green.* https://www.bloomberg.com/news/articles/2020-12-06/water-futures-to-start-trading-amid-growing-fears-of-scarcity?sref=w8mEqFdc

CNN (2020, October, 14). Flint water crisis fast facts. *CNN.* https://www.cnn.com/2016/03/04/us/flint-water-crisis-fast-facts/index.html

Coyle, K. J., & Von Susteren, L. (2012). The psychological effects of global warming on the United States: And why the U.S. mental health care system is not adequately prepared. https://nwf.org/~/media/PDFs/Global-Warming/Reports/Psych_effects_Climate_Change_Ex_Sum_3_23.ashx

Denchak, M. (2018, November 8). Flint water crisis: Everything you need to know. NRDC. https://www.nrdc.org/stories/flint-water-crisis-everything-you-need-know

D'Olimpio, L. (2021) Critical perspectivism: Educating for a moral response to media, *Journal of Moral Education, 50*(1), 92–103. doi: 10.1080/03057240.2020.1772213

Dufresne, T. (2019). *The democracy of suffering: Life on the edge of catastrophe, philosophy in the Anthropocene.* McGill-Queen's University Press.

Durham, M. G. (2018). Resignifying Alan Kurdi: News photographs, memes, and the ethics of embodied vulnerability. *Critical Studies in Media Communication, 35*(3), 240–258. https://doi.org/10.1080/15295036.2017.1408958

Franch, S. (2020). Global citizenship education: A new 'moral pedagogy' for the 21st century? *European Educational Research Journal, 19*(6), 506–524. doi: 10.1177/1474904120929103

Francis, Pope. Laudato Si': On care for our common home. 2015, http://www.vatican.va/content/francesco/en/encyclicals/documents/papa-francesco_20150524_enciclica-laudato-si.html

Guattari, F. (2008). *The three ecologies.* (I. Pindar, & P. Sutton, Trans.). Continuum.

Hoggett, P. (2019). Introduction. In P. Hoggett (Ed.), *Climate psychology: On indifference and disaster* (pp. 1–19). Palgrave.

Holden, E., Fonger, R., & Glenza, J. (2019, December 10). Revealed: Water company and city officials knew about Flint poison risk. The Guardian. https://www.theguardian.com/us-news/2019/dec/10/water-company-city-officials-knew-flint-lead-risk-emails-michigan-tap-water

Hoogsteder, L. M., Stams, J. M., Figge, G. J., Changoe, M. A., van Horn, K., Hendriks, J. E., & Wissink, J., I. B. (2015). A meta-analysis of the effectiveness of individually oriented cognitive behavioral treatment (CBT) for severe aggressive behavior in adolescents. *The Journal of Forensic Psychiatry & Psychology*, 26(1), 22–37. https://doi.org/10.1080/14789949.2014.971851

Human Rights Watch (2016, April 13). Make it safe: Canada's obligation to end the First Nations water crisis. Human Rights Watch. https://www.hrw.org/report/2016/06/07/make-it-safe/canadas-obligation-end-first-nations-water-crisis

IPCC (2018). Summary for policymakers. In: *Global warming of 1.5°C. An IPCC special report on the impacts of global warming of 1.5°C above pre-industrial levels and related global greenhouse gas emission pathways, in the context of strengthening the global response to the threat of climate change, sustainable development, and efforts to eradicate poverty* [Masson-Delmotte, V., P. Zhai, H.-O. Pörtner, D. Roberts, J. Skea, P. R. Shukla, A. Pirani, W. Moufouma-Okia, C. Péan, R. Pidcock, S. Connors, J. B. R. Matthews, Y. Chen, X. Zhou, M. I. Gomis, E. Lonnoy, T. Maycock, M. Tignor, and T. Waterfield (eds.)]. *World Meteorological Organization, Geneva, Switzerland*, 32.

IPCC (2021) Summary for Policymakers. In: *Climate Change 2021: The Physical Science Basis. Contribution of Working Group I to the Sixth Assessment Report of the Intergovernmental Panel on Climate Change* [Masson-Delmotte, V., P. Zhai, A. Pirani, S. L. Connors, C. Péan, S. Berger, N. Caud, Y. Chen, L. Goldfarb, M. I. Gomis, M. Huang, K. Leitzell, E. Lonnoy, J.B.R. Matthews, T. K. Maycock, T. Waterfield, O. Yelekçi, R. Yu, and B. Zhou (eds.)]. Cambridge University Press. In Press.

Janzen, M. [@janzen_m]. (2021, April 6). Preparing for @UM_Education grad class tonight. Interesting word counts in the @MBGov "BEST" report. [Tweet]. Twitter. https://twitter.com/janzen_m

Karoub, J. (2017, February 17). Commission: 'Systemic racism' at root of Flint water crisis. AP. https://apnews.com/article/df42de2ec4424193866467a2981ccb51

King, N., Bishop-Williams, K. E., Beauchamp, S., Ford, J. D., Berrang-Ford, L., Cunsolo, A., IHACC Research Team, & Harper, S. L. (2019). How do Canadian media report climate change impacts on health? A newspaper review. *Climatic Change*, 152(3), 581–596. https://doi.org/10.1007/s10584-018-2311-2

Kolbert, E. (2014). *The sixth extinction: An unnatural history*. Picador.

Kurth, A. E., (2017). Planetary health and the role of nursing: A call to action. *Journal of Nursing Scholarship*, 49(6), 598–605. https://doi.org/10.1111/jnu.12343

Klein, N. (2019). *On fire: The burning case for a green new Deal*. Alfred Knopf Canada.

Lonergan, M. D. (2020). Hard-on of darkness: Gore and shock websites as the dark tourism of digital space. *Porn Studies*, 7(4), 454–458. https://doi.org/10.1080/23268743.2020.1720524.

Lorenzoni, I., Nicholson-Cole, S., & Whitmarsh, L., (2007). Barriers perceived to engaging with climate change among the UK public and their policy implications. *Global Environmental Change. Part A, 17*(3–4), 445–459. https://doi.org/10.1016/j.gloenvcha.2007.01.004.

Manitoba. (2020). *Speech from the throne to open the third session of the 42nd legislative assembly of the province of Manitoba.* Winnipeg: Manitoba Government. https://www.gov.mb.ca/thronespeech/thronespeech-2020.html#educationandchildcare

Manitoba Organization of Faculty Associations (2021). *Manitoba PCs plagiarizing legislation from an American corporate lobby group.* https://mofa-fapum.mb.ca/issues/manitoba-pcs-plagiarizing-legislation-from-an-american-corporate-lobby-group/

Manitoba School Boards Association. (2018). Fire safety guide: Fire safety in Manitoba education facilities. http://www.mbschoolboards.ca/documents/services/riskManagement/studentSafety/Fire%20Safety%20in%20Educational%20Facilities%20-%20Guide%20for%20Educators.pdf

Martin, N. (2002, December 9). Wall street vultures are ready to get rich from water scarcity. TNR (The New Republic): Sold/Short. https://newrepublic.com/article/160484/wall-street-vultures-ready-get-rich-off-water-scarcity?s=09

McGrath, M. (2021, August 9). Climate change: IPCC report is 'code red for humanity.' BBC News. https://www.bbc.com/news/science-environment-58130705

McKibben, B. (2006). *The end of nature.* Random House Trade Paperbacks.

McKibben, B. (2019). *Falter: Has the human game begun to play itself out?* Henry Holt and Company.

Porter, J. J., & Evans, G. (2020). Unreported world: A critical analysis of UK newspaper coverage of post-disaster events. *The Geographical Journal, 186*(3), 327–338. https://doi.org/10.1111/geoj.12353

Purohit, K. (2019, November 5) As Debt Grows, More Indian Women Farmers Taking their Lives. https://www.aljazeera.com/indepth/features/farm-debt-grows-indian-women-farmers-lives-191023193523782.html

Russell, A. (2020, December 2). Indigenous service minister says Trudeau government won't end boil-water advisories by March 2021. Global News. https://globalnews.ca/news/7497223/indigenous-services-minister-says-trudeau-government-wont-end-boil-water-advisories-by-march-2021/

Stone, P., & Sharpley, R. (2008). Consuming dark tourism: A thanatological perspective. *Annals of Tourism Research, 35*(2), 574–595.

Svoboda, M. (2020, March 10). Media coverage of climate change in 2019 got bigger – and better. Yale Climate Solutions. https://yaleclimateconnections.org/2020/03/media-coverage-of-climate-change-in-2019-got-bigger-and-better/

Taubman, P. M. (2012). *Disavowed knowledge: Psychoanalysis, education, and teaching.* Routledge.

Thunberg, G. (2019, January 25). Our house in on fire. [Conference Session]. World Economic Forum. Davos, Switzerland. https://www.weforum.org/focus/davos-2019

Thunberg, G. (2019, January 25). 'Our house is on fire': Greta Thunberg, 16 urges leaders to act on climate. The Guardian. https://amp.theguardian.com/environment/2019/jan/25/our-house-is-on-fire-greta-thunberg16-urges-leaders-to-act-on-climate

Tschötschel, R., Schuck, A., & Wonneberger, A. (2020). Patterns of controversy and consensus in German, Canadian, and US online news on climate change.

Global Environmental Change, 60, 101957. doi: https://doi.org/10.1016/j.gloenvcha.2019.101957

van der Linden, S. L., Leiserowitz, A. A., Feinberg, G. D., & Maibach, E. W. (2015). The scientific consensus on climate change as a gateway belief: Experimental evidence. *PLoSOne 10* (2), e0118489–8. https://doi.org/10.1371/journal.pone.0118489

Van Susteren, L., & Colino, S. (2020). *Emotional inflammation: Discover your triggers and reclaim your equilibrium during anxious times.* Sounds True.

Wang, X., Thompson, D. K., Marshall, G. A., Tymstra, C., Carr, R., & Flannigan, M. D. (2015). Increasing frequency of extreme fire weather in Canada with climate change. *Climatic Change, 130*(4), 573–586. https://doi.org/10.1007/s10584-015-1375-5

Warburton, M. (2020, August 6) *Canada's last fully intact arctic ice shelf collapses.* Huffpost. https://www.huffpost.com/entry/canada-last-arctic-ice-shelf_n_5f2c9d69c5b64d7a55f0d6ec

Wilkins, C., Gobby, B., & Keddie, A. (2021). The neo-performative teacher: School reform, entrepreneurialism and the pursuit of educational equity. *British Journal of Educational Studies, 69*(1), 27–45. https://doi.org/10.1080/00071005.2020.1739621

Wray, B. (2019, May). *How climate change affects your mental health* [Video]. Ted Conferences. https://www.ted.com/talks/britt_wray_how_climate_change_affects_your_mental_health?language=en

Zhang, X., Flato, G., Kirchmeier-Young, M., Vincent, L., Wan, H., Wang, X., Rong, R., Fyfe, J., Li, G., & Kharin, V. V. (2019): Changes in temperature and precipitation across Canada; Chapter 4 *in* Bush, E. and Lemmen, D.S. (Eds.) *Canada's changing climate report.* Government of Canada, Ottawa, Ontario, pp. 112–193.

2 Education and the Visuality of the Anthropocene

I have felt a deep connection to birds for as long as I can remember. They connect me spiritually with my more-than-human relatives, and their unique songs are a prominent soundtrack in my life. When I am in their presence, I feel the tension from the work day leave my body and a wide awakeness take its place. Migratory birds mark the change in seasons in my world. It does not feel like Spring until I see the distinct V-flight pattern of Canadian Geese or hear them calling to each other as they fly over the house. Each year, these amazing birds return to the same region as their parents, often to the same nest. Their annual "return to the nest" speaks to me of the powerful influence formative relationships exert and why we often feel compelled to return home. These liminal waterfowl wonders spend as much time on land as they do in the water, and in this way, they invite an attunement to living and working in the in-between spaces in our lives. Over the hundreds of hours spent bird-watching, I have learned a great deal about the fragilities and strengths humans share with their feathered relatives.

Figure 2.1 A Pair of Geese in Flight

DOI: 10.4324/9781003024873-3

During a recent walk about the town, I encountered the in/famous Wild Turkeys of Brandon. The birds make their home in the southeastern part of the city, and they frequently generate conversation in the coffee shop, the school yard and in the media. There are some citizens in Brandon who complain the birds feed and roost on private property. The sentiment embedded in the complaints is, *if you don't pay the mortgage around here, get off my lawn*. In the past, The Brandon Police Service has even offered "tips" for people when they encounter these wild birds in "urban-rural interface zones" (CBC News, 2016). When I stopped to take photos of the male, I thought about the power relations enmeshed in the ways different Beings claim space in communities. While in the presence of the wild turkeys, I question what it means to live more relationally with our more-than-human relatives and problematize our attachments to property and ownership.

Figure 2.2 Gobbler Claims Space

In the Fall of 2021, I spent a frigid Sunday morning studying and photographing some common terns. Despite the bitterly cold wind, I became enthralled by the way the terns foraged over the open water. For hours, I watched them circle high above the lake and then drop straight out of the sky to pluck out a tiny fish. At one point, I traced the flight of one of the terns and noticed him carry a fish to a female tern floating in the water. I have learned since, as part of the courting process, male terns offer food to female terns to fortify a bond.

Education & Visuality of Anthropocene 35

Figure 2.3 Foraging for Fish

Terns are migration wonders; the average round trip of a common tern is approximately 35 000 kilometres each year. Watching these beautiful and buoyant fliers work with, and sometimes against, the wind and the waves, reminds us life and living are not self-enclosed. The wind and water are a part of the terns' relational field and it takes effort and dedication to the Other, to persevere when life's trajectories create distance or produce feelings of estrangement.

Figure 2.4 Cementing a Bond

Over the last two years, I found myself taking several photographs of flickers of rewilding. Tree branches thrust through broken windows of abandoned houses, prairie grasses swaying above formerly manicured lawns, a deer walking down a once busy thoroughfare quieted by COVID-19, the images became a poetics of possibility. On one level, the photographs are evidence of healing through ecological processes (Flyn, 2021) and on another, they are an anaesthetization of reclamation. Am I looking for a way to sit with the difficult knowledge, most Beings on the planet would be better off without us? Is this an attempt to find beauty in decay or some sort of eco-voyeurism? What I do know is many of the photographs I have taken speak to the long shadow of human interference after we have erased ourselves from the scene. Turning around and pointing one's lens towards the Other, is a way to question and confront what we leave behind in our wake.

Using a feeling-photography, I want to draw attention to climate precarity, biodiversity losses and other planetary boundaries exceeded by human activity in the wider context of educational research. My central goal in this chapter is to interrupt the pattern of ecological erasure in the field of educational research. I begin by describing a few impediments restricting our visuality and then I consider how photography, and narrative photography in particular, can illuminate the aesthetics of the Anthropocene. Moving against the dominant selfie-culture, I turn my lens towards Others to open a curriculum trajectory across different spaces and among different species. Feeling-photography is a way to, literally and figuratively, consider the values and emotionality embodied in particular frames of curriculum and pedagogy we have internalized.

What Makes It Harder to See

The failure to see the anthropogenic causes of the climate crisis cannot be reduced to a dearth of information. Global warming is a hyper-object (Morton, 2013), a phenomenon that is almost impossible to grasp because of its magnitude and how much it impacts every aspect of our lives. It crosses vast distances of time and space, and because the scale of climate change is enormous, it is incredibly difficult to hold a picture of it in our mind. A significant factor in the "everywhere and nowhere" status of climate change in education is the systemic datafication of learning. Datafication churns curricula into long lists of learning outcomes and it reduces pedagogy to deskilling recipes. In a system in which learning is dissolved into bits and bytes, there is limited emotional, intellectual, spiritual or embodied space left to view the connections between the grand bargain of continuous progress and the extraction of resources at the expense of our more-than-human relatives.

A *piecemeal epistemology* does not afford teachers and students much time to discover and understand the relationships among changes in land use, droughts, food shortages, designer seeds, pesticides and intensifying global production. No, a piecemeal epistemology obfuscates the real cost of your 99-cent hamburger! There is little time to linger upon the most pressing human questions. Instead, we continue to run, sweaty and tired, on an academic treadmill to nowhere.

Teachers, and teacher educators, are socialized to study, become an expert in the field and then profess what they know to their students but "…measured against the relative stability of the last epoch of life on Earth, the Anthropocene is a time of chaos and unpredictability, in which the frequency of (random, unpredictable) black swan events challenge the very meaning of what passes for normal" (Dufresne, 2019, p. xiii). That is to say, the field is writhing and much of the theory and practice in educational research disavows the rupture. Ecological erasure, a hyper focus on social mobility and the conflation of principles of inclusion with closing achievement gaps in performative systems, is distracting us to death. As we continue to lose thousands of our human and our more-than-human relatives to climate volatility, will education pay for the funeral, stand at the graveside and weep, bring cut flowers, join hands or pray?

In a field that is sometimes mired in happy affects, I do worry about striking the right chord of abjection. In a similar fashion to singer songwriter, Tom Waits, I do like "beautiful melodies telling you terrible things". It is not because I am more comfortable in spaces of moody pessimism, it is because half of the child and adolescent psychiatrists surveyed in England, 47 out of the 82 who replied, reported their patients suffer from environmental anxiety (Watts & Campbell, 2020). Fuelled by concerns for their future, and angered by the inaction of adults, students across the globe walk out of therapist's offices with their fists in the air. Time is of the essence, and young people have swallowed the ticking clock. Yet, in the field of education, we have yet to respond in any significant way, to the psychological harm eco-anxiety inflicts upon the young people we care for in our classrooms.

We live on shaky ground, and the curriculum-as-plan is insufficient when trying to examine complex, multi-perspectival and affectively charged dilemmas. A limiting factor in acknowledging the danger of living beyond Earth's carrying capacity has been the psychological defences people employ to protect themselves from difficult knowledge. Defences are discussed more fully in Chapters 4, 5 and 9. I argue in these chapters an evasion of the climate crisis has fostered epistemic mistrust. Uncertainty and mistrust proliferate among young people when their worries are dismissed or denied by significant adults in their lives. Educators could make an important difference in the emotional lives of their students if they acknowledged the severity of the situation, admitted adults should have acted sooner and worked in solidarity with young people to mitigate the damage.

I am moved by Claire Hogan, a 21-year-old climate activist, creator of Force of Nature, who said in an interview with *The Guardian*, "We're going to see massive, massive widespread climate crisis in every country around the world, so it's about developing the emotional resilience to carry on, but in a way that ignites really dramatic individual initiative" (Taylor & Murray, 2020). Exploring the affective dimensions of the climate crisis in the context of visual arts-based research, will not only give researchers and educators a chance to use what they learn to enhance the mental health and well-being of young people, but it will also open trajectories across schools and campuses to visualize and enact eco-compassionate living and learning. What is called for now is an expanded visuality in the field of educational research, one in which the human subject is decentred and no creature is pushed beyond the vanishing point.

Walking Delta Beach at Night

I often wander at the edge of the lake just after the sun ducks below the horizon. The vitality of eerily lit trees on the lakeshore and apparitions of distant sandbars conjure a ghostly research mise-en-scène for feeling-photography. My senses are heightened. The crack of twig under my shoe or the hoot of an owl brings about the uncanny even when I am treading over old ground. Walking in twilight, the everyday is somehow punctuated and aestheticized by a world-disclosing eeriness. Wagamese (2016) describes these moments of attunement as "…watch[ing] the universe shrug itself into wakefulness, as night surrenders slowly to day and shadow relinquishes itself to light. I watch this display and realize that the moon lives in the lining of my skin, the sun rises with my consciousness, and the earth thrums in the bottoms of my feet. Everywhere I go, I take that sense of wonder and mystery with me" (p. 90). Being sensory-struck is a weird encounter that haunts taken for granted subject/object demarcations and frames of reference. Often, I am sensory struck in the twilight and then compelled to take photographic sketches in response. It is in the moment when the sway of the willow branches shifts my gait or the second I am drawn to the stark contrast between a piece of driftwood against the moonlit sky. It is the decision to document a swirl of carp fish tails thrashing about in the channel because the sound of the splashing rippled down my spine. Linear notions of time, shift and bend in the echoes of aporetic subjects. Feeling-photography is a sensuous engagement with one's surroundings to situate oneself within the more-than-human world to "become [more] answerable for what [I] learn how to see" (Haraway, 1990, p. 190).

Education & Visuality of Anthropocene 39

Figure 2.5 Swirl of Carp at Twilight

Photography is psychoanalytic. Both the photographer and the analyst are always working with and against the preservation of affects, memorializing moments with the hope of returning someday to recover a relationship, feeling or an echo of identity in the present. The process facilitates the reading of oneself and Others back in time, which makes photographers and psychoanalysts proficient in the art of mourning. One of the most important books written about mourning and photography is Roland Barthes's, *Camera Lucida* (1981). Grieving his mother, Barthes went looking for her in old family photos. He describes the process as being "…alone in the apartment where she had died, looking at these pictures of my mother, one by one, under the lamp, gradually moving back in time with her, looking for the truth of the face I had loved" (p. 67). Through his examination of the photos, he made several important observations about the affective dimensions of photography.

Barthe introduces us to the *punctum,* a term encapsulating the "accident that pricks me" (p. 27) when one looks at a photograph. He describes the moment when an image crawls inside of you, forces you to look again and then re-emerges in the tiny hairs that stand up on the back of your neck. His poignant engagement with photography highlights the intersubjective nature of the artform. Even the prolific selfie maker thinks about the Other's judgment in the context of her relationship to a real or imagined audience. For its intersubjective situatedness (photographer, field, subjects, audience) and its capacity to facilitate mourning, feeling-photography is a worthy methodology to examine the frames of educational research within the aesthetics of the Anthropocene. Framing enters the act of photography at a very basic technical level.

The rectangular frame of the camera viewfinder, and its camera plane, was part of the design early in the history of photography. The intent of using a rectangle was to adjust the camera image so it resembled traditional art, the rectangular easel painting (Snyder & Allen, 1982, p. 68–69). The rectangle continues to shape the artform, but its origin story is long forgotten. I am drawn to feeling-photography as a way to rediscover the forgotten rectangles in educational research and education more broadly.

Narrative Photography in the Anthropocene

We tell stories to understand our situatedness in a wider field of relations. In our visual-centric lives, narrative photography plays a very important role in weaving those narratives. Sometimes referred to as staged photography, these images present a frozen story, often resembling a casual snapshot or a movie still (Barilleaux, 2016, p. 11). Figures 3.2, 3.3, 3.7, 3.8 and 3.10, in the next chapter are examples of staged photography. Evident in the first salted paper prints, narrative and drama thread through the entire history of the artform (Pauli, 2006). One of the most prominent historical examples is the *Self-portrait as a Drowned Man* (1840) created by Hippolyte Bayard to depict his suicide by drowning to criticize the French government for ignoring his role in the invention of photography (Pauli, 2006, pp. 13–14). In this iconic staged photograph, Bayard has his shirt off, he is slumped to one side, with his hands folded in his lap. The viewer sees his body from the waist up. A large straw hat hangs on the wall behind him to signal the arrest of the body. Amidst the murky and singed yellowish hues, Bayard creates a paradoxical encounter that is both proof of his existence and his erasure. Even the earliest days of staged photography made what was absent palpable.

In the late Victorian period, popular drawing room entertainment consisted of family members and guests dressing up in costumes and arranging themselves in frozen poses inspired by historical events, famous paintings or scenes from literature (Pauli, 2006, p. 16). Victorian photographers were not only characters in the scenes they depicted, but directors who made decisions about characters, composition, costumes and props. The artificiality of staged photography became less fashionable in the late 19th century and the early 20th century when handheld cameras and easily produced prints constructed the photographer as the earnest witness in journalism and scientific research. Less concerned with aesthetics and interpretive potential, photography became an important tool to disseminate the truth about human behaviour. For much of the 20th century, photography remained grounded in the pursuit of authenticity and truth. It was not until the late 1970s when artists began to use staged photography to challenge notions of photographic realism (Spencer, 2011).

Throughout history, photography and educational research have been identified with the evidential, viewed as technical processes to disseminate knowledge. An evocation of beauty or deliberations on the good life have been marginalized and characterized as dispensable luxuries or worse, interfering in epistemological debates. Consequently, learning cast in opposition to affect has made it very difficult for educational researchers to grapple with the sombre registers in the aesthetics of the Anthropocene. Photography's potency is that it illuminates trajectories between multiple sources of feelings. The photographed, the site, the viewer and the photographer, all exert a mutually influencing affective charge through the field of relations. Photography, then, is a powerful methodology through which the affective ecologies of the cultural, political and psychoanalytic realms are constituted and entangled.

Since every photograph is contingent, particularly when an image is pensive, oscillating feelings of kinship can move through the lens of the camera, producing fresh connections among materiality, fragility, dis/appearance and loss. Amid the spectacle of late-stage capitalism, research methodologies must be able to hold the living and the dead in the same space. Escalations in violence from white supremacist groups, massive losses in biodiversity, polluted air, rises in sea levels, drought, food shortages, extreme weather events and rising temperatures are all indicators that we need a language of mourning and moody pessimism in our work. We might do well to follow the trajectory of mourning through feeling-photography and psychoanalysis because there is no language in educational research to speak of the possibility of human extinction.

The theatricality of feeling-photography can hold the tension between the surreal and the real. Spectral figures (see Figures 3.1, 3.2, 3.8 and 3.10), melting fields (see Figures 2.6 and 3.7) the beauty of our more than human relatives (see Figures 2.1, 2.2, 2.3, 2.4 and 12.2) and finding the peculiar and evocative in the real (see Figures 2.5, 3.1, 3.4, 3.6 and 12.1), invite the maker and the viewer to engage with imaginative representations of self and Others. Playing with illusion, feeling-photography gestures towards imminent transformation in spaces of curriculum and pedagogy. To view the aesthetics of the Anthropocene, one must account for the contextual embeddedness of *looking* and attune to emotionally encoded visualization processes. Staged photographs are representations, the result of fixing ideas and giving meaning to one's lived experience. The images have material properties but they are always entangled within a scopic regime, meaning what is seen and how it is seen, are culturally constructed (Rose, 2016). Feeling-photography, as a methodology, focuses on who and what we are not attending to when we cast Others as unfamiliar objects. The more you make the traces of visuality conscious, the more capable you become at bearing witness to the d/evolutions in the web of relations sustaining your precious life.

The ubiquity of selfies and the proliferation of images on social media speak to the power of the image in the documentation and curation of our personal and professional identities. In June 2020, I gave a virtual keynote address to a group of high school art students. After my speech, the students and I spoke at length about how our Instagram eyes determine, to a large extent, what and how we see. We are living in a time in which the audience is thought of before the image comes into frame. For instance, when we go out in the world, we might subconsciously construct events, people, animals, plants and objects as future posts on social media. The students and I realized, when we pointed our cell phone cameras towards ourselves and used the place in which we stood as an ornamental backdrop to our consumer driven lives, it changed our relationship to the natural world in fundamental ways. Inspired by my conversation with the thoughtful art students in the spring of 2020, and mobilized by the pent-up energy from sheltering in place during the COVID-19 pandemic, the seeds of a photography project began to grow. In an attempt to allow the unconscious to have its way with me (Roth, 2016), I allowed myself to suspend concerns about perspective and composition and start by taking several inchoate photographs when I caught a glimpse of the monster-aesthetic of the Anthropocene.

The *photographic sketches* captured some of the awkward intersections between human activity and some of the wild places in Manitoba. Rough, sometimes haphazardly composed, the images evolved into emotional tracings and added layers to subsequent photographs which appear in the next chapter. I now conceive of these inchoate photos as flickers of movement, mood, tone or texture. The sketches mapped affective ruptures and opened new trajectories of recognition in wild places. As I reflected on the symbols and tones in the initial images, other objects, moods and locations for the staged photographs began to percolate in my mind.

Narrative photography co-constructs unconscious negotiations of highly charged concepts. It is a way to visualize one's theories of possibilities and problems by communicating one's inner life to the beholder. It begins with emotional compositing and ends with building a frame for unwieldy thoughts. The photographs in the next chapter contain emotional wounds and lingering reflections on my own positionality as an educational researcher in the Anthropocene. They represent a part of the affective matrices that entangle me, but I hope Others will bring their own wonderings to the content, perspective and imagery.

As a high school teacher, and then as a university professor, sometimes I have been left wanting when a colleague ends a speech or an article with, "Here's what teachers or principals ought to do..." in the absence of offering an image of praxis. For this reason, in the next chapter, I share part of an ongoing visual arts inquiry with you called, *A Prairie Elegy for the Discerning Consumer*. The text and the images are intended to create a liminal reflexive space to visualize the aesthetics of the Anthropocene. I welcome the tensions created by the entanglements among emotion, image and language. Images disclose themselves differently to each viewer (Eng, 2014) and so the photos are a *making with* the text and the viewer to make room for the unbidden. I hope the compositions elicit surprising fusions among the various concepts addressed throughout the book.

A Prairie Elegy for the Discerning Consumer

A Prairie Elegy for the Discerning Consumer resonates among the *ecological, social and mental registers* (Guattari, 2008) of curricula. As I curated the photographs and wrote the corresponding captions, some of what is distracting us to death in teacher education programs was drawn into sharper relief. Brewed in an alchemy of grief, longing and love, the images and text stretch down to the roots of living beyond the Earth's carrying capacity. My turn to narrative photography was a way to confront who and what I was not attending to when I cast Others as unfamiliar objects. Haraway (2003), reminds us that "...one cannot know the other or the self, but must ask in respect for all of time who and what are emerging in a relationship" (p. 50). I believe the more I make my limited visuality conscious, I can better trace where my care and concern goes, how it is expressed and where it lives among the three registers.

The interpretive spaces in the captions and images, problematize anthropocentrism by limiting the appearance of the human body and by centring objects and Others but, like all images, these ones are leaky. My theoretical wager is that visual interruptions, of and by the strange, produce a haunting vitality capable of disrupting unhealthy patterns that replicate the anthropocentric rigidity of what constitutes the terrain of research in education. Seizing upon the sensory rich mutability of digital photography, I look closely at the inextricable links among subjectivity and the relationships we cultivate with other Beings. Over the past five years, I have cycled through intense periods of climate grief. In search of the sociality of the senses to challenge inherited anthropocentric assumptions of subject–object relations, I composed images of estrangement when the fidelity to my own species was in play or when I recognized my eco-anxiety beginning to exert a paralyzing effect. Through the creation of the narrative photos, I was able to adopt more of a compassionate approach in relation to myself and to Others, particularly during periods of time when it was difficult to sleep. Part of the shift of being gentler in the face of monstrous feelings, is being more responsive to what becomes abundant instead of what is missing when uncertainties are in a state of play.

Narrative photography has made me more conscious of how I am tutored into my anthropocentric subjectivity by unmasking the latent content of my imagistic desires. The photographs may also produce a compounding effect when Others bring their own lived experiences to the images. In this way, my project is a pursuit that extends beyond my immediate intersubjective experiences to build a wider and wilder sociality. The most important parts of the process are the moments when I am estranged in once familiar surroundings. Once more a stranger at home, a new trajectory may "create – not destroy or undermine – other people's or other life forms' freedom to flourish" (Olvitt, 2017, p. 403).

In the intersections among visual storytelling, photography and drama which comprise the elements of narrative photography, I am given a form to represent the affective encounters that shaped (and continue to shape) the intensification of my relationship within an ailing natural world. Affects circulate as unconscious forces that compel bodies to respond within a field of relations (Seigworth & Gregg, 2010). In sensational encounters between bodies, bodies are changed, and in turn, bodies reshape encounters which make affects social, contextual and irreducible to individual subjects. Affects stand apart from the intentions of subjects because they leak beyond bodies in question and leave traces in the remains of past, present and future encounters. When an image stops you in your tracks, it is an invitation to connect with what deeply matters, even if the language to describe it as such is still on the way.

Admittedly, it can be terrifying *to look* at the Earth's ecosystems teetering on the edge of collapse. Morton (2013) compares the phenomenon to waking up inside an object, like a movie of being buried alive. Perpetually jolted between the ears, we find ourselves reading the last sentence of an apocalyptic fairy tale. But at the end of this grim story, the characters thrash about in a quaking bog, the instability of the surface of their lives catches up to them. The thick and spongy tenuousness of their social contracts waterlogged; they flail as the collective weight of their legs and toes break through the vegetation and into the frigid water beneath. The bog quake is dangerous, not because it sucks them in, they drown in the panic, stress and exhaustion that hangs heavy in their lives.

A Prairie Elegy for the Discerning Consumer asks why it is easier to imagine the end of the world than it is to work together to cultivate a more ecosophical approach to living. I am radically hopeful for what educational research can do in response to the question. Visual research methodologies can hold some of the tension between fatalism and hope within compelling experimental fictions. When the ghosts of the Anthropocene encounter the living, unconscious defences can be abstracted and then metabolized. Together, we can draw urgent attention to the destruction of long-evolving interdependencies and nurture our research affinities and curiosities to become response-able stewards of our more-than-human relatives. I invite other educational researchers to consider the Earth's decreasing habitability in the context of their work to bend curriculum and pedagogy towards compassion and mutual recognition and away from competitiveness, indignity and suffering.

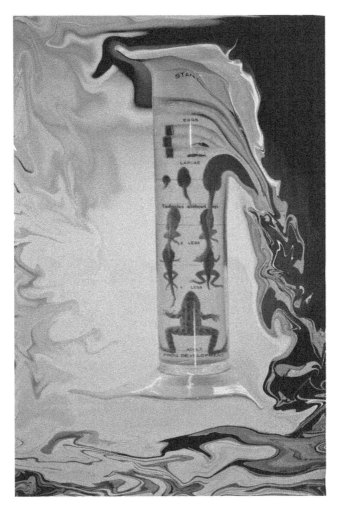

Figure 2.6 The Mise-En-Scène of the Anthropocene

References

Barilleaux, R. P. (2016). Introduction, In R. P. Barilleaux, A. Garza, G. J. Harris, & L. Soutter, (Eds.). *Telling tales: Contemporary narrative photography*. Mcnay Art Museum.

Barthes, R. (1981). *Camera lucida: Reflections on photography*. (R. Howard, Trans.). Hill and Wang. (Original work published 1980).

CBC News (2016, April 28). *Brandon's wild turkeys prompt advice from the police*. https://www.cbc.ca/news/canada/manitoba/wild-turkeys-brandon-police-tips-1.3557364

Dufresne, T. (2019). *The democracy of suffering: Life on the edge of catastrophe, philosophy in the Anthropocene*. McGill-Queen's University Press.

Eng, D. L. (2014). The feeling of photography, the feeling of kinship. In E. H. Brown & T. Phu (Eds.), *Feeling photography* (pp. 325–348). Duke University Press.

Flyn, C. (2021). *Islands of abandonment: Nature rebounding in the post-human landscape*. Viking.

Guattari, F. (2008). *The three ecologies*. (I. Pindar & P. Sutton, Trans.). Continuum.

Haraway, D. (1990). *Simians, cyborgs, and women: The reinvention of nature*. Routledge.

Haraway, D. (2003). *The companion species manifesto: Dogs, people, and significant otherness*. Prickly Paradigm Press.

Morton, T. (2013). *Hyperobjects: Philosophy and ecology after the end of the world*. University of Minnesota Press.

Olvitt, L. L. (2017). Education in the Anthropocene: Ethico-moral dimensions and critical realist openings. *Journal of Moral Education*, 46(4), 396–409. https://doi.org/10.1080/03057240.2017.1342613

Pauli, L. (2006). Setting the scene. In L. Pauli (Ed.), *Acting the part: Photography as theatre*. Merrell.

Rose, G. (2016). *Visual methodologies: An introduction to researching with visual materials*. Sage.

Roth, M. S. (2016). Why Freud still haunts us. In A. Harris, M. Kalb & S. Klebanoff (Eds.), *Ghosts in the consulting room* (pp. 120–124). Routledge.

Seigworth, G. J., & Gregg, M. (2010). An inventory of shimmers. In M. Gregg & J. Seigworth (Eds.), *The affect theory reader* (pp. 1–25). Duke University Press.

Snyder, J., & Allen, N. W. (1982). Photography, vision and representation. In T. Barrow, S. Armitage & W. Tydeman (Eds.), *Reading into photography*. University of New Mexico Press.

Spencer, S. (2011). *Visual research methods in the social sciences: Awakening visions*. Routledge.

Taylor, M., & Murray, J. (2020, February 10). 'Overwhelming and terrifying': the rise of climate anxiety. The Guardian. https://www.theguardian.com/environment/2020/feb/10/overwhelming-and-terrifying-impact-of-climate-crisis-on-mental-health

Wagamese, R. (2016). *Embers: One Ojibway's meditations*. Douglas & McIntyre.

Watts, J., & Campbell, D. (2020, November 20). Half of child psychiatrists surveyed say patients have environmental anxiety. The Guardian. https://www.theguardian.com/society/2020/nov/20/half-of-child-psychiatrists-surveyed-say-patients-have-environment-anxiety

3 A Prairie Elegy for the Discerning Consumer

In this chapter, I share ten digital photos (Figures 3.1, 3.2, 3.3, 3.4, 3.5, 3.6, 3.7, 3.8, 3.9 and 3.10) and the elongated captions from a larger visual arts project I am working on called *A Prairie Elegy for the Discerning Consumer*. The work is intended to open an interpretive space to examine the roots of the ecological crisis enmeshed in the field of educational research. More specifically, the work is an interrogation of the *leakiness of social-emotional necrosis*. The activities surrounding this wild and wilding inquiry shift between walking outside, staging concepts in photographs, research, writing, sculpting objects found outside (in a do no harm ethos), observing other creatures, daydreaming and noting compelling images in the aftermath and re/configuring the material world into megabytes. The images represent the connections I am making among the psychological, political and social dynamism of the retrenchments of late-stage capitalism at the end of the world. Some of the concepts have announced themselves in my head at night and resulted in staged photographs like Beak Wars (Figure 3.2), others like Taming Meme-i-mals (Figure 3.9) pressed themselves upon me while walking outside leading to novel connections. The project has allowed a multiplicity of sensations to bubble up from the unconscious. The work has entranced and alienated me at different stages of the process. Consequently, I understand this project as sur/real storytelling and a process to artfully attune to the world within me and around me.

I will not spend much time here making a case for arts-based research or what is more commonly referred to now as research-creation (Loveless & Manning, 2020) in education. I avoid a lengthy justification in this chapter because I discussed the importance of expanding one's visuality in the previous chapter; but more importantly, arts-based researchers have devoted too much time and page space justifying their existence in the academy. I am getting tired of being tired of this performative exercise. I start from the assumption that arts-based work is an integral component of learning to think disobediently. Furthermore, this type of creative work invites speculative ways of relating, which may help us to cultivate a more compassionate sociality as we come to understand our

DOI: 10.4324/9781003024873-4

48 *Prairie Elegy for Consumer*

disavowal in the field as a debilitating illusion. There are, however, a couple of things worth mentioning before you peruse the photographs. First, I chose a black, white and grey colour palette to invite a granular gaze, one that focuses on the contrast within and among the images. Second, in my attempt to speak about the ways in which the social, political and cultural tendrils of late-stage capitalism move through my body and the more-than-human-world, I intentionally left tiny time-shifting universes between the words accompanying each of the images. The gaps are spaces to imagine what alternative futures will look like if we choose to confront or continue to avoid the climate crisis in the field of education.

*****

Figure 3.1 The Weedy Tangles of Akiya

Going on luxury cruises made it difficult for Boomers to speak to their rightful heirs. You can't blame them. The reception is not great from the middle of the ocean. To apologize, Boomers bought and bequeathed coiffured headpieces woven from endangered prairie grasses to braid their children's fate with the Land. After meeting with their parent's lawyers, all the tiny sojourners were forced to migrate from the fields and forests to urban centers. While they walked, they wore elaborate masks to memorialize who they thought they were. At twilight, the minor witnesses would encounter strangers who shape-shifted into wildlife photographers. The poor man's photographer viewed the children as collectors' items to be posted on disaster tourism sites in the remote corners of the dark web. Upon their arrival, it was customary for the tiny townies to hide from the camera flashes in the bushes and watch the legacies dine out on the uncertainty of the times. And after the Great Ecological Collapse of 2060 (GEC), researchers recognized the inventiveness and resiliency on display during the child migration waves. In peer-reviewed papers, social scientists wrote nostalgically about why the masks the little warriors grew into, the ones with the papier mâché eagle beaks, became all the rage when they were all enraged in the wilder sociality of adulthood.

*****

Figure 3.2 Beak Wars

Right-wingers starved the system and then fed off political donors who could afford to buy the whole table. While the entrées were served, they commiserated about the bloated public education system and longed for the days when it was okay to give a kid a swat when he needed to learn how the real world worked. They patted the waiter on his ass when he agreed teachers were too damn soft nowadays. Even university professors cared more about their students' precious feelings rather than what they learned. By God, back in their day, they had to memorize the multiplication tables and smile while they did it. And for Christ sakes, why didn't somebody tell them when competition became a dirty word! It was bloody pathetic everyone received a trophy. What schools and universities needed was a little more beak on beak action and some accountability if they were going to compete in the global economy. No wonder [insert name of random country here] was eating their lunch.

*****

50 *Prairie Elegy for Consumer*

Figure 3.3 Harry is Hollow

    The serpents sizzled when God kissed the Gardener. The Aristocrats got the whole thing on tape (Scala Naturae gone wild!) and blackmailed the almighty into singeing the worms who writhed below the Gardener's feet. Nobody really kicked up a fuss because who is really going to miss snakes or earthworms? Admittedly, many tears were shed when lions, giraffes and elephants were carted off out of town, but you can't stop God when he's on a roll. In the South, they smelled what was cooking, but unfortunately, they were ghosted by Harry Hollow's surrogate wire mothers in the North. On Sundays, the richest Northerners broadcasted thoughts and prayers from the conductor's car each time the Robins raised a fuss about the earthworms. Unlike the Robins, however, people with credit card balances were moved by gold-plated prayers and well wishes. Consequently, each time a king or queen was derailed by paltry quarterly earnings, everyone in the yard signed a sympathy card for their Mother. The temporary workers, the ruffians who had "Fuck Harry Hollow" tattooed on their arms, sang songs for the Robins about maternal separation as they rode the funeral train to the next stop.

*****

Prairie Elegy for Consumer 51

Figure 3.4 Zombie Hot Wheels

When the world economy fizzled, Arts and Culture funding was immediately squeezed. Artists researched their own history of recycling to piece together some beauty in scraps. Freed from the confines of the gallery scene, revolutionary eco-artists used trashscapes as living canvases to visualize alternative futures. Through the cultivation of new assemblages of found objects, their recycled dreamscapes memorialized human creativity and offered young people radical hope. Activists and other creatives used the common spaces surrounding many of the pop-up artworks to nurture grassroots movements and to mobilize against the increased militarization of the police. Massive metal sculptures offered inspiration and protection to new generations of eco-culture warriors. One the most famous sites of resistance on the Canadian Prairies is pictured above. The sculpture was built by an artist called Xanex (2052–2092) and her apprentices. Her seminal work "Zombie Hot Wheels" became a rallying cry after the militarized zone expanded from Winnipeg to Thunder Bay in 2086. In addition to being a prolific protest metal artist, Xanex was widely respected for initiating a global network of teach-ins in which young people could learn about artists BGEC (Before the Great Ecological Collapse) who used their work to critique and organize against the structures and systems of climate apartheid.

*****

52 *Prairie Elegy for Consumer*

Figure 3.5 The Fragility of Butterfly Wings

On the first day of school, Ms X screen shared an image of a monarch butterfly. The chat box immediately filled with hearts and smiley face emojis. She told the students the monarch butterfly was a marvel of nature. Because the students had never seen a butterfly in person, she began the lesson by telling them butterflies had four wings – two forewings and two hindwings. Then she explained each butterfly wing consisted of two membranes that were covered with thousands and thousands of colourful scales and hairs. When she animated the photo, the students watched the butterfly's wings move up and down in a figure-eight pattern as it floated above the flower. She said they, too, would work as hard as the monarch butterfly once did during its annual migration. Their mouths dropped when they learned some monarchs flew over 4,000 kilometres to reach their winter homes. Ms X said the class would become strong and resourceful like the monarch butterfly and each student in her eyes was like a precious burst of colour on the butterfly's wing. Then she zoomed in on the edge of one of the butterfly's forewings. She asked the children to describe what they saw. The words "rip", "torn" and "hurt" filled the chat. Although the monarch was strong and resourceful, its wings could be fragile. Ms X said each one of them had a responsibility to care for other people in the class, and if anybody ever needed help, they could draw a figure-eight in the air with their pointer finger and someone in the class would send them a hug emoji and make sure they were okay.

*****

Figure 3.6 Mother and Calf Share a Secret in the Pasture

Cows were gentle creatures with complex feelings. They loved their babies and formed close friendships with other members of the herd. When family members or friends went missing or died, they grieved. After calves were six months old, they were taken off grass, put into backgrounding pens and forced to eat grain and corn. The diet sped up their lifespans and fattened them up so humans could marvel at the marbling in their steaks. Long before the GEC, cows whispered among themselves about the psychosocial cost of cheap fast food. Humanity's myopic focus on profit, efficiency, expansion and techno-fetishizing was literally killing them. Not only did the runoff from factory farms pollute rivers and lakes, animal agriculture was responsible for more greenhouse gases than all the world's transportation systems combined. Industrial farming would put them all out to pasture in the end. On the bright side, neo-Freudians finally proved the death drive was real.

*****

Figure 3.7 You Aint Nothin But a Slime Ball, Cryin All the Time

Hardy slime moulds, an amazing lineage of soil amoebas, made exotic colours and shapes that looked like rainbow popsicles (Newkirk & Stone, 2020). When social scientists sliced the rainbow-makers to extract the indigo, the slime moulds were justifiably incensed. In response to the experimenters' unprovoked acts of social impropriety, the slime moulds mobilized resistance by showing their captors they could reproduce with only 50% of their slime to work with. A savvy social scientist (the woman who came after et al. in the reference list) wrote a paper about it and exclaimed, "A slime mould was able to reproduce at 50% and was able to grow its indigo-less body toward a small piece of food at the opposite end of a maze. When the slime mould hit a dead end, it retracted its branches, retraced its steps and found another path". After her paper was published, her grad students coveted the high number of citations while missing the evolutionary genius on display. Using complex sets of signals, slime moulds employed elite-level ballet choreography that allowed 20,000 individuals to form a single blobby body. Humans had 50 years to get their act together before needing to move to cramped climate-controlled bio-zones, and they spent most of it bloodying their heads against the first partition of the maze.

*****

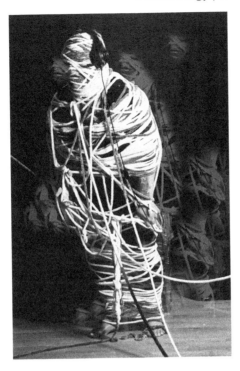

Figure 3.8 Digital Tethers

They were manic documentarians of their own lives, paparazzi with no history. It was customary for them to spend 5–7 hours a day curating their identities across multiple social media platforms. They shed their infantile narcissism in trillions of megabytes in the form of duckface selfies, photos of ornate sushi rolls and remixed dance videos. For decades, echoes of self-reference flooded the digital circulation system and diminished their capacity to think the thought of community. Disobedient thinking and negotiating difference were quelled in service of maintaining a peaceful coexistence within one's digital habitat. They became increasingly satisfied with being together alone. After the GEC, there were fewer hours available in the day to seek validation within the economy of likes. As wildfires raged and food shortages increased, more time was devoted to finding food, work and a safe place to live. But the owners of big tech companies were efficient digital colonizers. In response to the citizenry's new-found clickbait resistance, the Aristocrats of the digital world designed new algorithms to feed the public an alchemy of infotainment and insecurity. They increased the number of minutes people spent in front of screens by linking access to one's social media platforms directly to the economy of retweets and likes. Users who did not generate the requisite amount of online traffic were suspended from their platforms. New user agreements compelled people to upload copies of their digital selves over and over again, each time making their posts more salacious or violent to gain enough attention to maintain access. They had forgotten copies are doomed to degenerate. Digital echolocation disorders became ubiquitous. As the distance increased between users' social media echoes and what they deemed their real selves, loneliness and paranoia intensified.

*****

Figure 3.9 Taming Meme-i-mals

Insert your own meme [here].

*****

Figure 3.10 When Smoke Gets Under Your Eyelids

Covered in sweat, she pinches her thigh to make sure she is still alive. What did the shrink say to do again? Name things you see in the room. Say them out loud, girl! Pillow, red blanket with a small a hole in it, chipped paint on the ceiling, Ruby's squeaky ball, Ruby...She reaches for the glass of water on the nightstand. Her hands shake so much the glass falls to the floor. Broken glass on the floor, fan whirring in the bathroom, pink pajama shorts...In most of her dreams she smells smoke. She hasn't slept through the night in two years. Yellow sock on the floor, picture of Carlo on the dresser, blue laundry basket...In her dreams, there is always a fire burning. It's a terrifying moment when you realize your survival is random. She reaches for Ruby. She shouldn't look at her phone. Don't panic. Use your preplanned route. Move away from the wildfire, never toward it. Get back, get back, get back, get back, get back, get back. Don't leave before telling someone in uniform. If wildfires are burning, don't light any new fires. Keep in touch with absent household members. Call mom, call mom, call, mom, call mom, call mom. Don't let Ruby outside! Ensure your car is fueled and operational – a savior and a match on four wheels. She pinches her thigh again so hard her eyes water. A few minutes tick by. She wipes her face with her sleeve and gives thanks as the smoke begins to clear.

References

Loveless, N., & Manning, E. (2020). Research creation as interdisciplinary praxis. In N. Loveless (Ed.), *Knots and knowings: Methodologies and ecologies in research creation.* (pp. 211–220). University of Alberta Press.

Newkirk, I., & Stone, G. (2020). *Animalkind: Remarkable discoveries about animals and revolutionary new ways to show them compassion.* Simon & Schuster.

4 The Emotional Impact of Living in the Climate Crisis

Empirical evidence of the acute and chronic mental health effects of climate change has significantly increased over the last decade (Cunsolo et al., 2020). Beyond the community of climate scientists, more people are feeling disruptions to what was once the predictable cycles of their natural environments. Lingering feelings just under our skin warn us something is not quite right with the world. In July 2020, *climate goosebumps* rippled across my body twice within one week when the emergency siren sent our family to the basement because tornado clouds were spotted just outside the city. As I leaned against the washing machine holding onto to our shaking dog Maiv, I mumbled to my partner, "The thunderstorms in Manitoba are more intense now", and as the claps of thunder rattled the windows, one of our sons wondered out loud if "...tornado alley was moving north". *Weather talk isn't just weather talk anymore.* You, too, may have experienced the uncanny as you overheard a loved one mention, "The summers are so much hotter now" or maybe you can recall getting lost in your thoughts for a few seconds after your neighbours lamented the absence of monarch butterflies in their flower garden.

The United Nations predicts that by 2030, half of the people in the world will be living in places where there will be significant water shortages. As people wait in vain for a spring rain that never arrives, conflicts over water scarcity will increase in frequency and severity. Related food shortages, infectious diseases and wildfires will continue to impact the physical health of human beings. The massive changes to ecosystems have also caused more young adults to weigh their reproduction choices in relation to issues of environmental degradation. While it is true that socioeconomic factors continue to drive the decline of birth rates, more people are factoring climate change into their decisions about the number of children they have. In 2018, a survey conducted by Morning Consult for *The New York Times* asked respondents who indicated they "expected to have fewer children than they considered ideal", to identify the reasons for their choices. Thirty-three per cent of the 1,858

DOI: 10.4324/9781003024873-5

respondents, women and men ages 20–45, identified being worried about climate change as factor (Cain Miller, 2018).

Conceivable Futures is a grassroots organization that hosts a website, gathers testimonials and facilitates dialogue about the relationships between one's decision to have a child and the increasing disruptions to ordinary life caused by global warming. Part of the mission of the organization reads as follows:

> Some of us look to the future and can't imagine bringing children into such a hot, troubled world. For some of us, exposure to the fossil fuel industry has already jeopardized our health, or the health of our children. Parenthood has galvanized many of us towards greater action. For most of us, the threats to our reproductive freedom have been radicalizing. The climate impacts we see are unfolding during a time of increasing restrictions on reproductive self-determination and access to healthcare.
>
> (para. 2)

The network's facilitators recognize it is critical to surface and metabolize existential angst about having children within an invitational and trusting relational space. Lifton (2017) explains the importance of collectively making sense of the existential threat climate change poses:

> By confronting dire catastrophe and taking in the resulting death anxiety, even the possible death of our species, we make the swerve possible. That death anxiety, no longer avoided, becomes a stimulus for a continuous dynamic of awareness and potential action. In that way, the swerve creates a state of mind appropriate to the threat. And death anxiety becomes an animating force that both enhances and is kept in check by the swerve.
>
> (p. 103)

Young people, too, are awash in dire facts and projections of a bleak future. Helplessness and pervasive feelings of low-level stress continue to increase as talk of the planet being violated seeps further into their social media accounts and in the burgeoning cli-fi films and literature they consume. Schools ought to be places in which students, in the company of caring and thoughtful adults, talk openly about the impacts of climate change and what can be done to mitigate the challenges. To allow apocalyptic words and images to stay unprocessed in students' bodies in the place in which they spend so much of their time preparing for their future is intergenerational malpractice. One hundred years from now, it will seem inconceivable to teachers that on October 6, 2020, their ancestors saw footage of the first gigafire in modern history, a wildfire that burned over 1.03 million acres in the state of California (Kaur, 2020), and that it did not generate a mass mobilization in the United States and Canada to protect future classrooms of children from embers and smoke. When students ask themselves "with whom can we mourn the potential loss of a livable future", they should be able to turn to educators in their schools, colleges and universities for comfort.

In this chapter, I address the affective dimensions of the climate crisis, but more importantly, I explore some of the reasons we turn away from the extreme danger global warming presents and how the avoidance contributes to bringing about the horror we are trying so hard to evade. Psychoanalysts and climate psychologists (Hickman, 2020; Lertzman, 2015; Weintrobe, 2020) maintain the inaction is not due to a crisis of information. While it is true merchants of doubt have effectively peddled climate denials and in some countries like the United States, political activists have politicized climate change, politicization does not adequately explain the widespread inaction among leaders and policymakers in the field of public education. A lack of scientific information or the periodic deluge of misinformation on social media provides little clarity on the double-mindedness (Orange, 2017) at work that turns educators and policymakers away from the degradation of the Earth's ecosystems and the suffering of Others. It does not explain what is happening between our ears or in the gaps in curricula guides when those in the Global North continue to live off the profit from their colonialist ancestors. It does not illuminate the sources or escalating effects of guilt and shame exerted by compulsive consuming and a sense of entitlement to maintain harmful consumption practices and lifestyle choices.

The last gasps of late-stage capitalism continue to use the implements of standardization, deregulation, free market worship and de-professionalization to dig its own grave, which in turn has precipitated an infusion of "teaching by the numbers" (Taubman, 2012) in many school reform movements. As politicians fine-tune schemes to measure narrow learning objectives in scripted lesson plans, the complex intersections among the student's subjectivity and the teacher's subjectivity are ignored in a confluence of populist talking points and austerity measures. In turn, the mysterious unconscious, a bewildering and resonant affectivity in classrooms, is left out of conversations about mental health in schools to make room for a medical model that seeks to efficiently diagnose, treat and fix the messy emotions that leak out and interfere with the business of schooling.

Emotionally charged information often elicits defence mechanisms, so much so that a reorientation to ecosophical teaching and research will necessitate an *education in telluric emotions and a deeper understanding of the defences against them*. Admittedly, there are a number of barriers preventing collective and effective responses to the emotional pain caused by the climate crisis within the context of education. When it comes to mental health praxis, contextual analyses and therapeutic interventions typically centre on the lived experiences of individual children and their families. However, we know we do not exist within emotional vacuums, affected only by our own feelings. Psychological health is relational. What is called for is a greater awareness and study of *telluric emotions, emotions that are of the Earth, shared and mutually influencing.*

In December 2020, I attended a webinar organized by the Manhattan Institute of Psychoanalysis. The webinar series explored what relational psychoanalytic theory and practice could do to mitigate the increasing emotional toll of the climate crisis. As a psychoanalytically informed researcher in education, I was particularly interested in deepening my understanding about the role of anxiety that, both unconsciously and consciously, makes it harder for a more ecosophical worldview in education to emerge. During the webinar, I had the great fortune to listen to the eminent psychiatrist, Robert Jay Lifton. He spoke about the parallels among what he learned from the survivors of the nuclear bomb in Hiroshima (1982), the climate crisis, the witnessing professional and the Nazi doctors in World War II who propagated malignant normality. Lifton's work on malignant normality, psychic numbing and the witnessing professional illuminates some of the reasons why the climate crisis operates as a cipher in the field of education.

Psychic numbing (Lifton, 1982) describes an inability to respond rationally when confronted by the threat of extinction. The numbing is caused not by a traumatic event, but from a crisis of meaning (Norgaard, 2011, p. 4). Projections about the dire impacts of the climate crisis add up to an existential threat to the continuation of human existence. While the land is scorched, the honeybees die, prairie grasses choke on herbicide and the tree branches singe, teachers continue to live their professional lives in the same way all the while knowing life could be extinguished from its source sooner than one can bear to imagine. Consequently, as educators become more cognizant of the ways the climate crisis disturbs the mental health of children in their care, and as teachers become increasingly conscious of how global warming will wreak havoc on the habitability of the planet, psychic numbing may be one of the contributing factors to the psychosocial paralysis in the field. If the aim of one's profession is to prepare children for the future and life as we know it has been put into question, then it is understandable that the very thought of teaching produces a numbing effect when one experiences existential angst about the precariousness of the future. Psychological freezing, an evolutionary response to life-threatening events, becomes problematic when it binds a community of practitioners to a vow of silence. More specifically, when a helpful psychological reflex at the level of the individual is entrenched in a professional discourse, it fosters a malignant normality of silence which has to be named and confronted. More concretely, when professional groups repeatedly withdraw from ecocide to avoid psychological pain, they become ineffectual witnessing professionals whose silence may account for some of the missing references to climatological and ecological precarity in curricula, lesson plans, budgets, policy, strategic plans and classrooms.

Climate change crosses vast distances of space and time. Before the industrial revolution, CO_2 levels in the atmosphere were 270 ppm, and now, CO_2 levels exceed 400 ppm. Climate scientists may rock in bed over the increase but it is difficult for most other humans to measure the urgency to change in parts per million. Unfortunately, a lot of the damage done by human activity such as the burning of fossil fuels is exerted at the microscopic level. It is hard for the mother who cradles her wheezing child in the emergency room to point at the urban smog that snatches her baby's breath away. When we catch a glimpse of the more-than-human world biting back, normalcy bias, a state of thinking that underestimates the degree of threat that a disaster will have on us personally, is another way people retreat from emotional discomfort. There are many people in the Global North who have not been chased from their homes by wildfires or droughts and many who have not connected their children's inhaler prescriptions to air pollution. From a place of a psychosocial blindness, they scroll past images of exhausted faces in Red Cross tents on their way to booking airplane tickets for the family's Christmas vacation.

The widespread detachment from the affective dimensions of climate change in the field of education is often exerted through a discursive *regularization of happy affects*. The phenomenon repeatedly surfaces when a critical mass of teachers, students or administrators speak out against harmful structures and systems in education that contribute to the degradation of ecosystems. Over the years, a number of articles and books have sprung from the tendrils of school improvement discourses (Fullan, 2007; Hall & Hord, 2001) to diagnose sources of resistance and the requisite treatments. Many educational leadership texts in this vein diagnose those who criticize the assumptions and aims of school improvement processes as naysayers, or even saboteurs. Collective utterances about multisystem ecological critiques are quickly deemed gratuitously gloomy, political, and antithetical to progress or most commonly as an individual's unwillingness to accept *what we do, we do in the best interest of students*. To keep the professional peace, resistors swallow doses of hyponymy medicine which produces two harmful social side effects, one of the most pernicious being an attachment of the meaning of words like *worker, competitor, consumer* and *21st-century learner* to the student body and the other, a severance of school improvement initiatives from the wider web of life.

As ecosophical researchers, we must weave telluric emotions into our research stories to give language to the intense feelings attached to the disappearance of the muskox in Manitoba, the decline of the Barrens Willow in Newfoundland and the third mass bleaching event in the Great Barrier Reef. In other words, we must collectively recognize all our research stories are staged against the emotionally charged backdrop of ecological devastation. To that end, our questions, methods and dissemination strategies need to challenge the erasure patterns that obfuscate eco-anxiety and climate grief. One pathway a researcher can take to do this work is to deconstruct the cumulative effect of euphemistic language or what Stibbe (2021, p. 80) refers to as *purr-words* in curricula guides, policy decisions, standardized tests and lesson plans that gloss over our embeddedness in ecological degradation while severing students' emotional lives using neoliberal machinations. Wallin (2017) suggests that a more pessimistic dialogical approach in educational research might disrupt the culture of happy affects in the field:

> Pessimism might constitute a new disposition for reassessing the ideals of progress and optimism that continue to regulate pedagogical expression and research within the ambit of affective capitalism and its circuits of interminable productivity and semiosis.
>
> (p. 1108)

We cannot allow euphemisms or disingenuous peppy progress talk in the form of empty signifiers to reify the systems and structures of capitalism. As I discuss at length in the next chapter, I am strongly in favour of opening space for monstrous feelings and moody pessimism in the field of education. Happy affects are preventing a sober accounting of the precarity of teaching in the Anthropocene. For the most part, educators force themselves to smile while the brutal reality of an ecological crisis presses closer. To keep the uncomfortable feelings at bay in classrooms, chapter books like *The Hunger Games* (Collins, 2008), *The Maze Runner* (Dashner, 2010) and *Agnes at the End of the World* (McWilliams, 2020) are devoured. Literature circles and novel studies have become socially permissible metabolizations of images of children in the throes of an environmental apocalypse. Fantasies of young heroic figures who carve out a life for themselves amid an inhospitable world soothe and entertain. We know extreme weather events, poor air quality, heatwaves, rising sea levels and mass migrations will exert a psychological toll and make life more difficult, so we collectively transfer debilitating angst into fictional heroines like Katniss Everdeen in *The Hunger Games*. In Katniss's world, the totalitarian Capital selects a boy and a girl from each of the 12 districts of Panem (North America in the future) to participate in an annual televised fight to the death. Katniss survives by using her training, intellect and physical prowess to become our winner. Is sharing Katniss Everdeen's story how we get our grandchildren ready?

Telluric Emotions and Defences Against the Difficult Knowledge of the Climate Crisis

What is perplexing, particularly at this time in human history, is there is scant discussion in educational research about the creation of holding environments for students' feelings of climate dread. A short time ago, a friend of mine who is a high school teacher pointed to the irony of the situation. She observed students engaged in discussions about climate volatility and lamented, "Some of them are exhibiting a weird combination of being horrified and numb at the same time, and we're not doing enough to support them". To step into the bad object world, to become a proper holding environment for young people's existential climate angst, teachers and educational researchers must collectively shake themselves awake. We are not poorly written characters in a cli-fi film noir. Climate change is not an other-worldly spectre. It is possible to open space in classrooms and lecture halls for a relational psychological state, one appropriate to the threat, in which fear and sadness can begin to operate as an animating force for ecosophical teaching and learning.

As a way to open a small space for this type of inquiry, I am currently working with young people to build an eco-arts community of practice to better understand how they are experiencing climate anxiety and grief. I want to know more about what their worries are, how they are making sense of their feelings and what role an eco-arts community of practice can play in helping young people to make sense of their complex emotions. As a result of this work, I have learned there are too few concepts, frames or stories in education to speak of emotions in relational terms. In the first place, emotions are often subsumed in datafication processes, and if they are addressed, the experiences of individual humans are the focus, to the exclusion of social emotional states.

To delve into the study of telluric emotions, it is worth considering the source and manifestation of shared climate anxiety. The most destructive form of projection is the projection of anxiety onto one's environment. Mauss-Hanke (2013) calls it *terra cremata*, "the fear of re-entering psychic territories that had once been experienced as something most threatening and were therefore 'burned down' (from the Latin *cremare* – to burn) – that is, they were psychically destroyed" (p. 54). In relation to climate change, she uses the concept in reference to "the inner 'land' of an existential vulnerability and dependence – a terribly frightening and traumatic experience we all went through as babies (though of course at very different levels, frequencies and intensities) and which we overcame by developing our autonomy in order to never again be so vulnerable and dependent" (p. 54). We know at a cellular level our lives depend on the Earth. With each climate rupture, whether it is felt in one's community or dreams, the collective unconscious reels because the Earth, the home of all homes, is getting sicker and becoming less dependable. In response to the trauma, a social splitting occurs, opening a psychic void. The gap is then reinforced by an ideological machine that operates in service of a shared delusion to reduce the scale of the climate breakdown in the collective psyche.

Over 50 years ago, climate scientists warned burning fossil fuels and the associated carbon emissions would continue to warm the planet resulting in significant harm to the Earth's ecosystems. Rather than heed the warnings, Canadians continued to invest in car-centric communities, build pipelines through First Nations territories and order more trinkets through amazon.ca. According to Lotringer (1997), as cited in Cantz (2018, p. 12), social dysfunctions and "...disasters may highlight the massive insecurities that lie beneath the surface of an otherwise well protected cultural exterior" (p. 58). When shame flushes the cheeks of those who read about the 600 Amazon warehouse workers in Brampton, Ontario, who contracted COVID-19 (Mojtehedzadeh, 2021) or when the majority of Canadians ignore their brothers and sisters in the Neskantaga First Nation who have been deprived of safe drinking water for over 25 years (Stefanovich, 2020), it animates shame and guilt in the collective unconscious. Why? It is immoral to block the suffering stranger's attempts to impress themselves on one's self-satisfied life (Orange, 2011) and as "life enhancing entanglements disappear from our landscapes, ghosts take their place" (Gan et al., 2017, G2).

When we ignore the suffering of others, it exerts a dehumanizing effect on the stranger and the withholding-witness. I conceive of the *withholding-witness* as a person who is unable to address the pain of the Other in order to maintain a conscious or unconscious grip on the asymmetrical power relations in a given encounter. Narratives valorizing autonomy, hyper-individualism, the conflation of consumption with success and the commodification of other Beings in service of human ends cultivate the proliferation of withholding-witnesses, a defining social pathology of our time. Drawing from Levinas, Orange (2017) asserts people will often defend against their feelings of guilt or shame in a doer-done-to (Benjamin, 2018) encounter by blaming strangers for their own suffering. When we resort to this kind of psychological defence, for example, pathologizing the behaviour of individuals and denying the existence of structural violence, we enter the territory of moral and ethical obscenity. Orange goes on to warn it is impossible for groups and individuals to escape unconscious feelings of guilt caused by a dismissal of the suffering stranger. Rooted in formative encounters with caregivers, feelings of shame produce caustic emotional side effects, the most pertinent in the context of this work being the undercurrent of eco-anxiety that pulses through the social ecosystem.

Eco-anxiety is a relatively new term in academic, psychoanalytic and media circles to describe heightened emotional, psychological or embodied distress in response to the harmful effects of climate change. In 2017, the American Psychological Association published a report linking the impact of climate change to psychological distress and referenced the term *eco-anxiety* as the chronic fear of environmental doom (American Psychological Association, 2017). While the term is described as a "new" phenomenon in the media, for Indigenous people, eco-anxiety precipitated by land loss and climate precarity is not new phenomenon. That said, analyses of eco-anxiety, like all other social issues, cuts against the intersecting lines of colonialism, gender, race and class.

Historically, psychoanalysts have placed little emphasis on the relationship between planetary health and mental health. The world of treatment was constituted within the four walls of the consulting room. One of the first analysts to inquire into the relationship between ecosystem degradation and anxiety was Harold Searles (1972). He laid important groundwork for educational researchers to inquire into the psychosocial implications of eco-anxiety. Through his work, we can better understand the eerie observation of my friend who noticed her students appeared simultaneously numb and horrified when speaking about climate precarity:

> We live today at a time when we must save the real world or we shall use it as the instrument for destroying us all. I think that the greatest danger lies neither in the hydrogen bomb itself nor in the more slowly lethal effect of pollution from our overall technology. The greatest danger lies in the fact that the world is in a state as to evoke our earliest anxieties and at the same time to offer the delusional "promise," the actually deadly promise of assuaging these anxieties, effacing them, by fully externalizing and reifying our most primitive conflicts that produce these anxieties. In the pull upon us to become omnipotently free of human conflict, we are in danger of bringing about our extinction.
>
> (Searles, 1972, pg. 373)

Apocalyptic anxiety brews an alchemy of horror and numbness. It emerges when the permanence of life on Earth is thrown into question. People recoil when their "illusion of apocalyptic exceptionalism" (Cantz, 2018, p. 17) is threatened. Unlike the young activists whose protests are fuelled by their anger and worry about government fecklessness, there are far more young people across the world who retreat and swallow their existential angst. The subsequent psychological retreat motivates further evasions and destructive thinking, activating primitive anxieties and defences such as splitting, denial and fatalism. Living with apocalyptic anxiety has serious consequences. It can result in poor reality testing and reductions in proportional thinking. Further, it turns anxious energy, a vital response to potential threats to one's survival, into a state of being in which an individual's relatedness to their environment is severely distorted and undermined. The field of educational research can no longer evade the climate change dilemma. The widespread repression across the field has already caused intergenerational harm. If we understand at least part of our work within an ecosophical orientation, we must address the psychological injuries young people are suffering and develop new methodologies to account for the affective ecologies of the Anthropocene. Research in education ought to identify and trace sympoietic practices (Haraway, 2017) in educative encounters that assist young people in carving out a life upon the emotional landscapes of a damaged planet. These sympoietic, or *making-with pedagogies of affect*, requires an understanding of the fits and starts of malattunement and mutual recognition in nurseries and classrooms.

Psychological conflict exists within an intersubjective context, one that includes the more-than-human world. In relational psychoanalytic terms, the seeds of trauma form if caregivers respond with consistent malattunement to a child's affective state (Stolorow et al., 1987). When a child's emotional state becomes unbearable, the seeds of trauma take root in the absence of a relational home. Stolorow and Socarides (1984/1985, p. 106), as cited in Stolorow (2011), assert it is the "absence of steady, attuned, responsiveness to the child's affect states [that]…leads to significant derailments of optimal affect integration and to a propensity to dissociate or disavow affective reactions" (p. 26). In other words, to process overwhelming experiences, children require a space for affect integration or their emotional wounds will fester.

Imagine what it must be like for young people to witness the wilful destruction of the Earth, the holding environment of all holding environments, while many of the significant adults in their lives (parents, caregivers, teachers, spiritual advisors) dismiss their concerns, placate them or worse, deny their reality. A worried child needs to be addressed by a receptive other when extinction nightmares violently rock the cradle of a dysregulated social psyche. We urgently need to open more spaces around dinner tables and in classrooms for compassionate recognition so children can move towards the otherness of the future with an increased sense of security. Because many adults have not responded to children's cries with either enough honesty or urgency, eco-trauma has been embodied, encoded and passed on, leading to "whole clinics full of children with climate anxiety" (L. Vansusteren, personal communication, December 5, 2020). It is time to force the question, what is going on between the ears of adults (caregivers, teachers, instructors, spiritual advisors etc.) who profess a love for other people's children when they evade young people's (and their own) worries about the climate crisis? Weintrobe (2013) identifies three common defences caregivers use to avoid difficult knowledge. Her work is instructive to try and make sense of the widespread disavowal of shared anxiety associated with the intergenerational impacts of the ecological crisis.

One common psychological defence is the inflation one's presence and opinion to eclipse the subjectivity of the Other (Weintrobe, 2013). I describe these moments as *ego swelling*. It might sound like an instructor being sarcastic to a student during a class when he perceived his expertise was being publicly challenged by the student. In the wider public discourse, it might sound like radio talk show hosts calling young eco-protesters snowflakes. These types of encounters are often loaded with narcissistic signals, conscious and unconscious messages sent to others when we expect to be accommodated while claiming a position of exemption for ourselves. In most cases, ego swelling is a momentary defence against anxiety. It results in the denial of another person's subjectivity and a demand for idealization at the expense of mutual recognition. In the context of the climate emergency, puffing oneself up might sound like dismissing an environmental activist by making ir(rational) claims about how saviour technologies will rescue us in the end.

Another common defence against the pain associated with troubling information is projection (Weintrobe, 2013). Projection involves displacing one's anxiety onto others. During a recent grad class, I overloaded students with facts and figures about climate volatility. Even when I noticed three students get smaller inside their individual video conference frames by moving their chairs back from their webcams, I was driven to pressurize the learning space. After class, I recognized I had been projecting my own anxiety onto the students. Earlier in the day, I read a particularly gut-wrenching article on the climate-induced changes in the life of the Atlantic Ocean. I needed their hearts to race along with mine, so I delivered the requisite emotional jolts through IPCC projections to feel the beats.

A third psychological defence against the difficult knowledge of the climate crisis is denial. Denial is a way to defend against an overwhelming sense of fear by pretending it does not exist. It is an attempt to protect oneself from the pain that would result if we accepted the reality of our situation. Weintrobe (2013) writes about the differences among three different forms of denial people use to defend against the difficult knowledge of the climate emergency. She defines *denialism* (p. 7) as a campaign of misinformation about climate change that is funded by ideological and commercial interests. *Negation* (p. 7) is saying something *is* when it *is not*. It is often our first response when reality is too much to process. That said, negation is often a temporary defence and a signal that someone has begun to mourn a significant loss. A more pernicious form of denial is disavowal (p. 7). To disavow something is to know it and to not know it at the same time. Disavowal can be more difficult to work with because it requires some aspects of reality to be accepted but the significance of the loss is minimized. Disavowal can become an organized system in one's metal processes in which inner peace is maintained by perpetually denying the destructiveness of our actions. Disavowal silences the critical voices in our heads and leads us to behave in ways which bring about the terrible situation we are trying so hard to escape.

After one of my undergrad classes this year, a student mentioned he watched a documentary about the prolific environmental activist David Attenborough. He was moved by the stunning images in the film and horrified about the extent to which deforestation was causing harm to animals. He could not understand how the millions of people watching Netflix could have this information and then do nothing about it. At one point during the conversation, he burst, "Don't you think we are being insane"! His exclamation contains one of the most significant educational research questions of Anthropocene. How can we be so calm in the face of the unprecedented danger posed by the ecological crisis? Later in our conversation, he refers to what he interprets as the silence of his high school science teachers. He wondered aloud if they knew about the disappearing Amazon rainforest. His anger, fear and disappointment were palpable, triggering my own anxiety about the world my children and future students will inherit. I fell silent for long stretches of time during the conversation. I struggled to find good enough words. Emotional paralysis in shared silences are dialogic shadows of undigested panic, loss, and guilt, and they make it difficult to contain existential dread.

Downloading Our Apocalyptic Anxiety

Searles (1972) made the case that we deal with the speculative extinction of the human species by projecting our non-human conscious desires for omnipotence and control onto technological objects. In the science fiction movie, *Archive* (Rothery, 2020), the central character George communicates with his dead wife Jules through the Archive, a tall, ominous-looking device resembling a tomb. He spends the tortured moments between communicating with his dead wife in an abandoned military base working on several artificial intelligence (AI) projects. The first robot George creates, J1, is boxy and cumbersome. Limited in her ability to communicate, J1 responds to George as a well-behaved toddler would. The second robot, J2, has frightening dreams, converses with George, challenges his worldview and actively cares for J1. J2 also expresses jealous feelings when George shifts his time and focus to building J3, revealing to the audience J2 experiences human emotions. We come to learn J1 and J2 are George's first attempts at creating a lifelike container for the remnants of his dead wife's consciousness stored in the Archive.

Through transplanting recovered elements of his wife's essence into J3, the film interrogates the traumatizing effects of death anxiety. If you can tolerate some of the gendered fantasy tropes infecting parts of the movie (J2 covets the body of J3; J2 kills herself because she is jealous of

George's obsession with J3; George builds J3 a humanoid body that is sleek, slender and stereotypically attractive), there are several moments in the film to consider affectivity and trauma in the liminal space between life and death. The AI revenants, as fantasies of omnipotence, symbolize human desire for immortalization through technological advancement. But these three figures of estrangement become much more. In a plot twist at the end of the film, the audience learns it is George's consciousness uploaded in the Archive, not his wife Jules. Revealed as metaphysical illusions in George's subconscious, J1, J2 and J3 open a space to question what counts as a life. If J2 is capable of nurturing J1, challenging George and of dreaming, should her existence not count as an end in itself? In the last part of the film, J2 walks into the lake, killing herself as an unsettling answer to the question. The film presses on the contingencies of existence and contemporary evasions of the irreversibility of death. When we look through the lens of the Archive, inaction in the face of the climate crisis is not equanimity or a lack of compelling information (Heath & Heath, 2010); the repetition of our destructive behaviour (J1, J2, J3...) points to the ways shared un/conscious trauma shatters the absolutisms we build our lives around.

Climate change forces us to confront the loss of our shared home. There is no escape for George, for you, for students, for me, if the Earth as a dwelling place is going to disappear. Indeed, we are collectively experiencing more fear, despair, helplessness, grief and trauma in connection to the Earth's degradation (Davenport, 2017; Hamilton, 2019; Lertzman, 2015). But despite the daily news reports and a deluge of images on social media of severe weather events, flooding, droughts and looming food shortages, the affective dimensions of teaching and learning in the midst of the climate crisis are muted, if not disavowed. Teachers, leaders and researchers must account for the climate crisis in educational praxis because eco-anxiety, grief and depression will continue to increase among children in the years to come. We need an analytic attitude to explore what it looks like in educational contexts to bear witness to emotional discomfort, face our finitude head-on and share useful knowledge to bring about the mitigation of emotional pain. I suggest a turn to culture-making, experimental storytelling and visual arts activism may aid in the containment of monstrous feelings about climate change and alternatively, elicit fresh images of a more humane and generative future for us all.

References

American Psychiatric Association. (2013). *Diagnostic and statistical manual of mental disorders* (5th ed.). https://doi.org/10.1176/appi.books.9780890425596

Benjamin, J. (2018). *Beyond doer and done to: Recognition theory, intersubjectivity and the third*. Routledge.

Cain Miller, C. (2018, July 5). Americans are having fewer babies. They told us why. *The New York Times.* https://www.nytimes.com/2018/07/05/upshot/americans-are-having-fewer-babies-they-told-us-why.html

Cantz, P. (2018). Apocalyptic exceptionalism and existential particularity: The rise in popularity of dystopian myths and immortal "other. In E. R. Severson & D. M. Goodman (Eds.), *Memories and monsters: Psychology, trauma and narrative.* (pp. 11–22). Routledge.

Collins, S. (2008). *The hunger games.* Scholastic Press.

Conceivable Futures. (2021, February, 12). Mission. https://conceivablefuture.org/mission

Cunsolo, A., Borish, D., Harper, S. L., Snook, J., Shiwak, I., Wood, M., & The Herd Caribou Project Steering C. (2020). "You can never replace the caribou": Inuit experiences of ecological grief from caribou declines. *American Imago,* 77(1), 31–59. https://doi.org/10.1353/aim.2020.0002

Dashner, J. (2010). *The maze runner.* Delacorte Press.

Davenport, L. (2017). *Emotional resiliency in the era of climate change: A clinician's guide.* Jessica Kingsley Publishers.

Fullan, M. (2007). *The new meaning of educational change.* (4th ed.). Teachers College Press.

Gan, E., Tsing, A., Swanson, H., & Bubandt, N. (2017). Haunted landscapes of the Anthropocene. In E. Gan, A. Tsing, H. Swanson & N. Bubandt (Eds.), *Arts of living on a damaged planet: Ghosts.* (pp. G1–G14). University of Minnesota Press.

Hall, G., & Hord, S. (2001). *Implementing change: Patterns, principles, and potholes.* Allyn and Bacon.

Hamilton, C. (2013). What history can teach us about climate change. In S. Weintrobe (Ed.), *Engaging with climate change: Psychoanalytic and interdisciplinary perspectives.* (pp. 16–32). Routledge.

Haraway, D. (2017). Symbiogenesis, sympoiesis, and art science activisms for staying with the trouble. In A. Tsing, H. Swanson, E. Gan & N. Bubandt (Eds.), *Monsters: Arts of living on a damaged planet.* (pp. M35–M50). University of Minnesota Press.

Heath, C., & Heath, D. (2010). *Switch: How to change things when change is hard.* Crown Business.

Hickman, C. (2020). We need to (find a way to) talk about ... eco-anxiety. *Journal of Social Work Practice,* 34(4), 411–424. https://doi.org/10.1080/02650533.2020.1844166

Kaur, H. (2020, October 6). California fire is now a 'gigafire,' a rare designation for a blaze that burns at least a million acres. https://www.cnn.com/2020/10/06/us/gigafire-california-august-complex-trnd/index.html

Lertzman, R. (2015). *Environmental melancholia: Psychoanalytic dimensions of engagement.* Routledge.

Lifton, R. J. (1982). *Indefensible weapons: The political and psychological case against nuclearism.* Basic Books.

Lifton, R. J. (2017). *The climate swerve reflections on mind, Hope and survival.* The New Press.

Mauss-Hanke, A. (2013). Discussion: The difficult problem of anxiety when thinking about climate change. In S. Weintrobe (Ed.), *Engaging with climate change: Psychoanalytic and interdisciplinary perspectives.* (pp. 52–55). Routledge.

McWilliams, K. (2020). *Agnes at the end of the world*. Hachette Book Group.

Mojtehedzadeh, S. (2021, March 21). More than 600 Amazon workers in Brampton got COVID-Why were so few reported to the province? Toronto Star. https://www.thestar.com/news/gta/2021/03/21/more-than-600-amazon-workers-in-brampton-got-covid-19-why-were-so-few-reported-to-the-province.html

Norgaard, K. M. (2011). *Living in denial: Climate change, emotions and everyday life*. The MIT Press.

Orange, D. (2011). *The suffering stranger: Hermeneutics for everyday clinical practice*. Routledge.

Orange, D. (2017). *Climate crisis, psychoanalysis, and radical ethics*. Routledge.

Rothery, G. (Director). (2020). *Archive*. [film]. Independent; Hero Squared; Head Gear Films; Lipsync; Metrol Technology; Quickfire Films; Untapped; XYZ Films.

Searles, H. F. (1972). Unconscious processes in relation to the environmental crisis. *The Psychoanalytic Review*, 59(3), 361–374.

Stefanovich, O. (2020, December 17). After evacuation twice over tainted water, the Neskantaga residents plan their return home. CBC News. https://www.cbc.ca/news/politics/neskantaga-plans-return-home-water-crisis-1.5840308

Stibbe, A. (2021). *Ecolinguistics: Language ecology and the stories we live by*. (2nd ed.). Edition. Routledge.

Stolorow, R. D. (2011). *World, affectivity, trauma: Heidegger and post-cartesian psychoanalysis*. Routledge.

Stolorow, R. D., Brandchaft, B., & Atwood, G. E. (1987). *Psychoanalytic treatment: An intersubjective approach*. Analytic Press.

Taubman, P. M. (2012). *Disavowed knowledge: Psychoanalysis, education, and teaching*. New York: Routledge.

Wallin, J. (2017). Pedagogy on the brink of the post-anthropocene. *Educational Philosophy and Theory*, 49(11). p. 1099–1111. doi: 10.1080/00131857.2016.1163246

Weintrobe, S. (2013). The difficult problem of anxiety in thinking about climate change. In S. Weintrobe (Ed.), *Engaging with climate change: Psychoanalytic and interdisciplinary perspectives*. (pp. 33–47). Routledge.

Weintrobe, S. (2020). Moral injury, the culture of uncare and the climate bubble. *Journal of Social Work Practice*, 34(4), 351–362. https://doi.org/10.1080/02650533.2020.1844167

5 Anthropocentric Fantasies Entangled in the Pain of Eco-Grief

I walked down to the water to take some photographs of the constellation of shore birds gathering at the edge of the Delta Marsh. I was delighted to encounter a pouch of pelicans and a boisterous committee of terns. Ten minutes into the walk, my son Vanya joined me and we headed east along the shore. Off in the distance, we saw a little boy in red and white shorts jumping around waving his arms in the air. It looked like he was bouncing on top of one of many sandbars revealed by the warm Southwind. As we got closer, we noticed he was looking up at the birds circling overhead as he danced around in a circle. When he took notice of us, he stopped jumping and waved us over. He asked my son, "Do you wanna see something cool *and* gross"? The invitation was hard to refuse. We followed him to the other end of the sandbar, bent down and tracked his index finger to the head of a dead carp fish. The little boy was delighted to discover a fly now sat on top of the fish's eyeball. He queried, "Do you want me to touch the fly"? I don't think I wanted him to because I quickly interjected, "Were you just dancing and waving at the birds"? He replied, "I don't want the birds to eat him. His mom might be looking for him". At this point, the little boy's father joined us on the sandbar. After we exchanged some weather talk and his son updated him on the birds, the fly, the carp and the fish's mom, the father took his son's hand and smiled, "Let's get back. We don't want those [carp fish] here anyway. They're ruining the marsh".

In the invocation of *Islands of Abandonment*, Flyn quotes neuroscientist David Eagleman who proposes we have three deaths. The first death occurs when the body stops working, the second death happens when our bodies are cremated or buried and the third death is "that moment, sometime in the future, when your name is spoken for the last time" (Eagleman, 2009, as cited in Flyn, 2021, p. 9). In light of Eagleman's assertion, we experience three deaths. What can we make of the young boy who interrupted the second death of the carp? The encounter not only touches on the themes of fantasy, loss and mourning, it directs our attention to the values enmeshed in the time elapsing between the second and third death. For the boy's father, the carp's third death is built in to

DOI: 10.4324/9781003024873-6

the second. "We don't want those [carp fish] here anyway". The fish's life may have little value to him because the carp are "ruining the marsh". Alternatively, the boy's fantasy of the mother carp searching for her son momentarily circumvents the third death. To the boy, the carp's life has value because the fish has a mother who misses him. In waving the birds away, he understands himself in relation to the life of the carp and the carp's family.

To delve deeper into the connections embedded in the encounter, we must also see it staged against the backdrop of one of the world's largest freshwater wetlands located at the southern end of Lake Manitoba. In the 1960s, duck hunters sounded an alarm at Delta Marsh because they noticed a dramatic drop in the number of ducks in the area. Turbid waters, less pond weeds to hold the sediment in place and a decrease in the aquatic vegetation which provided food resources for diving waterfowl were cause for concern. Biologists pointed to human interference in the lake's stochasticity after the provincial government built the Fairford dam to regulate lake levels. In addition, they identified the Common Carp, a large bottom-dwelling "food fish" introduced to Manitoba over 100 years ago as the other main culprit. Carps use their large mouths to burrow into the sediment to feed on aquatic plants and tiny invertebrates. Weighing between 5 and 30 pounds, these turbidity tornadoes have been labelled by waterfowl biologists as disruptive members of the marsh community. Carp, fly, carp mother, diving duck, dam, sediment, boy, sandbar, pond weed, hunter, researcher, water – all connected.

When I saw the father grab his son's hand, I recalled my mom yelling during a dandelion funeral for a frog that occurred too close to the road. I connect the rigidity of the father's body as he listened to his son's worries about the carp's mother to the quizzical look on my mom's face as she brought the frog's funeral her seven-year-old daughter was officiating to an abrupt end. It makes me wonder about the moments when I figuratively grabbed my students' hands and pulled them away from making sense of their own worries and losses. Most importantly, the narrative abundant in the heartbeats of disenfranchised grief is a reminder that classrooms are full of little and large losses, many of which go unrecognized.

Grief

Grief is the term we use to try and describe the constellation of unwieldly emotions Beings experience after a significant loss. The introjection of loss requires one to work through telluric and tumultuous feelings like guilt, fear, anger, despair and longing. I use the word Beings instead of people in this context to force the inclusion of the more-than-human world in discussions of loss. Humans often minimize and discount the sadness other creatures experience. Consequently, when I speak of grieving in the context of this work, I understand it as a *telluric emotional state, one that is of the Earth, shared and mutually influencing.*

The Climate Psychology Alliance (2021) adapted Worden's (1983) work on grieving to describe what it looks like to embrace or reject the tasks of mourning in the context of climate change. To embrace the work of grief involves accepting a loss, working through excruciating emotions, adjusting to a new interpersonal field, developing a new sense of one's identity and finding a place to contain the loss. Rejecting the task of grief manifests as denying the facts and irreversibility of the loss, idealizing what was lost, bargaining, numbing, becoming bitter, feeling helpless or turning away from loved ones and the vitality of life. Climate grief can emerge when we experience physical ecological losses, the loss of environmental knowledge and we can also grieve anticipated losses (Cunsolo et al., 2020).

The philosopher Glenn Albrecht coined the neologism *solastalgia* to describe a form of emotional or existential distress caused by your changing environment. Albrecht (2019) defines solastalgia as:

> ...the pain or distress caused by the ongoing loss of solace and the sense of desolation connected to the present state of one's home or territory. It is the existential and lived experience of negative environmental change, manifest as an attack on one's sense of place. It is characteristically a chronic condition, tied to the gradual erosion of identity created by the sense of belonging to a particular loved place and a feeling of distress, or psychological desolation, about its unwanted transformation.
>
> (p. 39)

He contrasts the meaning of solastalgia with the spatially bound concept of nostalgia. He says "...in direct contrast to the dislocated spatial dimensions of traditionally defined nostalgia, solastalgia is the homesickness you have when you are still located within your home environment" (p. 39). Albrecht's work is instructive because climate change activates emotional registers capable of traversing vast distances of space and time. For example, one might have been awash with grief sitting in a classroom in Saskatoon, Canada, after learning 180 million birds were harmed in the bushfires that seared the Australian land and sky from June 2019 to February 2020 (Vernick, 2020). An experience with eco-grief is not only a recognition of the Earth as the home of all homes, it suggests that the compassionate witnessing of grief is not rooted in space and time. One can relate to a suffering stranger, another Being from a different place or period and be vicariously saddened or provoked by existential distress.

A more personal illustration of solastalgia, and how it can provoke the turn to compassionate witnessing, is the artwork tattooed on my arms. Part of my body has become a memorial to the extinct butterflies. As butterflies (and other Beings) are tattooed on my body, I open a space for me and others who may see the tattoos on my arms in person or view them via photographs to grieve the disappearance of butterflies across the planet (Agrawal, 2019). Taking my own body as a point of departure, I mark the disappearances, mourn and memorialize the extinction so my *sorrow is processed as a life-enriching embodied vulnerability*. When unprocessed grief is swallowed whole, ghostly reverberations of melancholy can fester and interfere with one's capacity to engage fully in the world. I know this from my research and personal experiences with eco-grief. There are beautiful and life-affirming feelings entangled with the sources of our collective grief. We grieve what we love and when we mourn, the process eventually makes room for joy and radical hope. If we do not mark our heartaches, the ghosts of the Anthropocene will haunt.

Ghosts of the Anthropocene are figures of rebellion and recalcitrance (del Pilar Blanco & Peeren, 2013). They call into question what is placed outside of knowing, that is, what is excluded from the language of loss. If a loss cannot be acknowledged as such, inexpressible mourning erects a psychic crypt. Its creation process is referred to as incorporation (Abraham & Torok, 1972). More specifically, incorporation describes a refusal to acknowledge the full impact of a loss in the wider sociality of subjectivity because if it was recognized as such, it would effectively transform the ones who have experienced the loss. In relation to the climate crisis, the horror of the sixth mass extinction is entombed and society's acknowledgement becomes a block of reality (disavowal), and reality is thus defined as what is evaded, obfuscated and masked. The reality of the ecological crisis is a secret to be guarded (climate denial, greenwashing, fetishizing technology...) as a mechanism of preservative repression:

> The crypt marks a definitive place in the [society's psychic] topography. It is neither the dynamic unconscious nor the ego of introjections. Rather, it is the enclave between the two; a kind of artificial unconscious lodged in the very midst of the ego.
>
> (Abraham & Torok, 1971, p. 157)

Here lies the reason why a psychoanalytically informed approach to educational research is integral. Psychoanalysis has been historically concerned with the ways in which figurative ghosts can be laid to rest as ancestors (Harris et al., 2016). Psychoanalytically informed educational research could lay open the endocryptic identifications in the aims, processes and practices in the field that cohere the figurative ghosts of late-stage capitalism in the social realm. An integral aspect of this difficult work is to acknowledge the trauma of ecocide (Stolorow, 2011) and to give language to those losses before students engage in preservative repression. Marc Schlossman's inquiry called *Extinction* is an example of an exercise of public mourning that honours and gives language to such losses.

For over ten years, Marc Schlossman photographed several specimens of extinct and endangered species in the zoology and botany collections of The Field Museum of Natural History in Chicago (Heinz, 2017). He began to capture the series of photographs he calls *Extinction*, (Schlossman, 2017a) because in his words he "wanted to say something about what is happening in an age in which human activity dominates the environment on an unprecedented scale". He goes on to say:

> We are now stewards of all other species and we abuse the responsibility. Our actions contribute to the accelerating loss of biodiversity. Man's ability to extract raw materials and produce consumer goods, many of which biodegrade very slowly, is harming the biosphere. Science advances at a faster rate than ever before. We prize our technological capabilities, including a belief that somehow our tools will solve any problem, but current technology cannot address the fact that our understanding of the consequences of our actions within complex ecosystems is less than our knowledge of an ecosystem's individual components.
>
> (para. 2)

One of the photos that continues to haunt me is his photo of the Hawksbill Turtle (Scholossman, 2017b). Set against a black backdrop, the bleached remains appear as elongated bony skeletal fingers that reach towards the viewer. If you move your cursor over the image it magnifies a section of the photo to reveal numbers that have been written in black marker on the remains. To me, the numbers represent the black market that profits from turning their tortoiseshells into trinkets and bobbles driving them ever closer to extinction. To explore Schlossman's photographic archive is to engage in an act of mediated public mourning. His photographs have not only deepened my understanding of how photography can create a relational space of public pedagogy, they provoke me to think more deeply about the traumatizing forces of disenfranchised grief.

Acknowledging the eco-grief of young people peels away some of the disavowals and the motivational posters on classroom bulletin boards to reveal anthropocentrism, social mobility and hyper-individualism as some of the structures that reify the psychic climate crypt. The laissez-faire attitude of adults regarding the eco-grief young people have internalized, can shake young people's sense of themselves and cause them to lose faith in the safety of their environment. False reflections of social reality that do not account for visible threats to humanity's survival corrupt the internal psychic landscapes of young people. To put it plainly, if a group of teachers knew the school caught fire and instead of ushering children to safety, they shouted "We're fairly sure you'll get out in time" as they ran out of the school, we would describe them as mad and their actions as wicked.

As it relates to the climate crisis, disenfranchised grief is embodied in the denials that lie in opposition to the climate change losses that many young people have internalized. Greenwashing, dismissals and placations can shake young people's sense of themselves or cause them to lose faith in the safety of their environment. The minimization of visible threats by significant adults in children's lives *is* and *will be* a corrupting force in the internal psychic landscapes of young people. The deafening silence about climate precarity turns young people's dependency on adults for support in times of tumult to an exercise in running through a hall of mirrors. Unable to ground themselves, they spin back and forth from normalcy to fear, as feelings of insecurity mount across the fragmented reflections of reality. To that end, facing the climate crisis in spaces of curriculum and pedagogy can interrupt the disenfranchisement of climate grief, and in the process make adults emotionally legible again. Furthermore, there are transformational feelings beneath the surface of grieving. When we mourn, it releases energy that can be redirected to the stewardship of our more-than-human relatives.

Internalized feelings of guilt and fear about global warming have caused many well-meaning adults to remake children into miniature saviours. Young activists like Greta Thunberg have been vaulted into the global spotlight as a testament to one's belief the children will "get out in time". Boomers borrow against the lives their great-grandchildren might have lived to build pipelines to fuel the sports utility vehicles they drive to go to the megastore to buy exotic fruit in December. *Good things happen to those who never give up on their dreams. You can be anything you want to be, as long as you put your mind to it.* Happy talk of the future may be a good way for adults to mitigate their own psychological pain, but when the gap between a young person's conscious reality and the fictions circulated by significant adults in their lives becomes too great, young people may displace their disappointments and anger onto themselves and Others in the future to protect the illusions that framed their young lives. Happy talk about the future can be pernicious to climate-aware children because it makes adults emotionally illegible.

Psychologists who write about the grieving process often cite building dialogic communities as an essential part of the healing process (Davenport, 2017). Our collective grief about oceans of plastic, clear cutting the Earth's lungs in the Amazon Rainforest and uncontrolled Alberta wildfires could be acknowledged and discussed more openly in classroom settings. This would require teachers to attune to expressions of eco-grief and to become more reflexive and resilient in the presence of *enactments* in dialogic encounters. Enactments occur when depth and nuance collapse in a conversation, when at least one of the participants assumes her assessment of the Other, event or idea in question is true and that it cannot possibly mean anything else. In educative encounters, enactments prevent the thoughtful questioning of assumptions and reinforce conventional wisdom. Conversely, *emergence* allows the thoughts and emotions of others to be moved into a playful space. In order to open safe and compassionate spaces in classrooms and lecture halls to process eco-grief, fostering mutual recognition must become a communication artform those in the profession and those who study the profession value and practice throughout their careers.

Only when we acknowledge the precarity of the time we find ourselves in the company of Others will we be able to achieve a productive tension among acknowledging young people's despair about their futures and maintaining the energy and optimism required to fight for our future. Striking a balance in precarious times is arduous but essential work. It begins with teachers understanding the classroom space as a holding environment (Winnicott, 1971/2005), and to do that, the first step is to acknowledge the messy feelings (anger, fear, hope, anxiety, grief) that come from knowing students are living in what can sometimes feel like the opening scene from a cli-fi movie. It requires admissions of responsibility and actively troubling shared fantasies of dominance and control over the wider environment.

Tracings of anthropocentric fantasies and attempts to mitigate loss can be found in Freud's influential deliberations on the construction of modern civilization. In *Civilization and its Discontents* (1929/2018), he asserts man must subjugate elements of the natural world for human ends and reifies the status of technology because of its protective factors.

> We recognize that a country has attained a high level of civilization when we find that everything in it that can be helpful in exploiting the earth for man's benefit and in protecting him against nature – everything, in short that is useful to him– is cultivated and effectively protected. In such a country, the course of rivers which threaten to overflow their banks is regulated, their waters guided through canals to places where they are needed. The soil is industriously cultivated and planted with the vegetation suited to it; and the mineral wealth is brought up assiduously from the depths and wrought into the implements and utensils that are required.
>
> (Freud, 1929, p. 36)

Freud tells us nature (something humans are separate from) is something to fear. Therefore, we are justified in our attempts to *bend the river to our will*. The image of water overflowing its banks taps into my own vulnerabilities, fantasies of omnipotence and desire for control.

When I was 23, my community flooded. In 1997, a winter storm brought a massive amount of snow to Manitoba in early April. As much as 50 centimetres of snow fell in the Red River valley in a 24-hour period on April 5 (Hoye, 2019). The level of the Red River was already high and when the snow from the blizzard melted quickly, the water had no place to go. As the river rose, my dad used a bright orange wooden stake to mark how far the water had encroached each day. When my sister or I expressed we felt exhausted, worried or sore during one of the sandbagging marathons, he would point to the wooden stake and say, "Just do what you can". After the sun set, we would habitually pull our lawn chairs together and set sandbagging goals for the next day. Every evening, as the sun set, my mom would serve cheese sandwiches and hot cups of coffee, and then repeat how the water looked to her like it was "swallowing the prairie landscape". My dad would add, "This whole place sounds different. I think the birds have gone".

Close to 8,000 soldiers from the Canadian Armed Forces were brought in to help with sandbagging and the evacuation of approximately 25,000 people, turning some communities into ghost towns. Although the provincial government discussed the impacts of the flood predominantly in financial terms (roughly CAD 500 million in Manitoba), the psychological impacts were profound. Mouldy drywall and rotting wood represented the material challenges of the flood, but even today, the RCMP officer's declaration that it was time to evacuate still echo in my family's memories. Weeks after we were forced to leave our home, my parents waited anxiously in a small hotel room for updates on the status of the house. They left precious childhood drawings and family photos soaking inside keepsake boxes. There was no way to regulate or protect against the affective dimensions of the Flood of 1997.

The mental health impacts of global warming will be global, cumulative and harsh. In Manitoba, the Canadian province I live in, we will likely face earlier and more severe changes to our climate compared to many other places in the world (Government of Manitoba, 2021). More severe storms, higher risks of flooding and periods of drought in farming communities are part of our future. More significantly, the number of people who will eventually become part of the climate diaspora in the Global South will be beyond our comprehension. The displacement of millions of people will lead to protracted civil unrest and push social systems like health care and education beyond capacity for extended periods of time. Increases in mood disorders, suicide, post-traumatic stress disorder, acute stress reactions, sleep disturbances and substance abuse will be some of the profound consequences of climate instability (Palinkas et al., 2020).

The mental health burdens on school systems will become so great that many of the responsibilities that used to be delegated to the school guidance counsellor, social worker or vice principal will be delegated to teachers. Consequently, it would be prudent for all teacher education and graduate programs of study to include courses and training that focus on mental health and well-being in climate-precarious times. More importantly, it is critical for mental health research and practice to broaden its scope to include telluric emotional trauma. Trauma as well as depression, grief, and anxiety are social phenomena, and teachers are in a unique and powerful position to build the capacity of students to collectively navigate emotional turbulence.

A significant part of the preceding discussion may have felt to you as too pessimistic, but we need to be able to muster the courage to despair (Sontag, 2007) in order to fight against eco-injustice. The courage to mourn is evidence of a deeply felt connection to the vitality of life and it provides sustenance as we process what we have lost in order to forge the relationships we need to survive in precarious times. My eco-grief and yours is evidence of the interconnectedness of our shared fate. I have not given up on you, me, our students or this work. It is possible to engage in transdisciplinary inquiries about grieving processes amid the ecological crisis in ways that validate, challenge and inspire. As an example of such a process, I invite you to take a tour of Dismaland.

Dismaland

Dismaland was a pop-up art exhibition of an apocalyptic theme park in a deserted seaside resort billed as "The UK's most disappointing new visitor attraction" (Jobson, 2015). It was the disobedient brainchild of Banksy, a prolific street artist and political activist based in the United Kingdom. Banksy's political commentary combines satire, other forms of bleak humour and graffiti to generate biting commentary on social issues. The official brochure from Dismaland characterized the event as a chance to embrace mindless escapism:

> Are you looking for an alternative to the soulless sugar-coated banality of the average family day out? Or just somewhere cheaper. Then this is the place for you—a chaotic new world where you can escape from mindless escapism. Instead of a burger stall, we have a museum. In place of a gift shop we have a library, well, we have a gift shop as well.
> Bring the whole family to come and enjoy the latest addition to our chronic leisure surplus—a bemusement park. A theme park whose big theme is: theme parks should have bigger themes...

Dismaland included the provocative artworks of over 50 artists who tackled the themes of environmental apocalypse, anti-immigration, consumerism, militarized policing and vapid celebrity culture. As an illustration, loss of freedom was poignantly on display in the artwork called the Museum of Cruel Objects, a bus filled with objects intended to inflict harm, curated by Gavin Grindon (Swirsky, 2015).

Dismaland's anti-capitalist vision is an outstanding example of an aesthetic deconstruction of the *stories we live by* (Stibbe, 2021). Young people in schools, colleges and universities ought to encounter the likes of Banksy, but more importantly, they should learn to develop their own critiques of the stories they have inherited, particularly the stories reifying colonial, capitalist and patriarchal frameworks. If given some space to compose and give language to their own counter narratives, stories about the real costs of fossil fuel extraction, for example, they might bring new holistic concepts of health, happiness and success into the world. While it may not be feasible to actualize a Dismaland-scale artwork of political resistance in schools, educators could design creative eco-justice inquiry projects in response to local community eco-injustices. It is within these types of critical inquiries that young people can compose and continually revise an ecosophical approach to living and learning.

If we continue to amplify happy affects in the field of education at the expense of introjecting eco-grief, we risk reinforcing the epistemological inadequacies that perpetuate the deadening discourses of education that got us into this mess. A consideration of eco-grief in the context of ecosophical education and research will not only expand our capacity for reverence amid the more-than-human world, it will enhance our ability to process intense feelings. Mourning facilitates the introjection of loss to make room for other feelings such as joy, passion and hope. Moreover, when we mourn, it recharges our emotional energy and orients us to what makes life worth living. If we are prepared to face our losses with courage and compassion, Schlossman's *Extinction*, the Flood of 1997 and Dismaland are examples of generative interruptions to the dangerous confluences of dominance, happy affects and disavowals. To be sure, climate change is transforming who we are and who we can become in the presence of Others. A relational turn in how we conceptualize and respond to the moody pessimism in education, a state of being ripe for accessing collective patterns of meaning production, holds some potential for enhancing psychic resourcefulness when the field fractures. This emotionally charged work will require educational researchers to engage with young people, amplify their voices, bear the difficulty of their emotions, respond to their questions with compassion and urgency and then work in solidarity in pursuit of an ecosophical orientation to living.

References

Abraham, N., & Torok, M. (1971). The topography of reality: Sketching a metapsychology of secrets. In Abraham, N., & Torok, M. (1994). *The shell and the kernel: Renewals of psychoanalysis*. (pp. 157–164). (Vol. 1). (N. T. Rand, Trans.). University of Chicago.

Abraham, N., & Torok, M. (1972). Mourning or melancholia: Introjection versus incorporation. In Abraham, N., & Torok, M. (1994). *The shell and the kernel: Renewals of psychoanalysis.* (pp. 125–137). (Vol. 1). (N. T. Rand, Trans.). University of Chicago.

Agrawal, A. A. (2019). Advances in understanding the long-term population decline of monarch butterflies. *Proceedings of the National Academy of Sciences of the United States of America*, 166(17). 8093–8095. https://doi.org/10.1073/pnas.1903409116

Albrecht, G. A. (2019). *Earth emotions" new words for a new world.* Cornell University Press.

Climate Psychology Alliance. (2021, April 23). *Handbook: Grief.* https://climatepsychologyalliance.org/handbook/321-grief

Cunsolo, A., Harper, S. L., Minor, K., Hayes, K., Williams, K. G., & Howard, C. (2020). Ecological grief and anxiety: The start of a healthy response to climate change? *Planetary Health*, 4, e261–e263. doi: https://doi.org/10.1016/S2542-5196(20)30144-3

Davenport, L. (2017). *Emotional resiliency in the era of climate change: A clinician's guide.* Jessica Kingsley Publishers.

del Pilar Blanco, M., & Peeren, E. (2013). Introduction: Conceptualizing spectralities. In M. del Pilar Blanco & E. Peeren (Eds.), *The spectralities reader: Ghosts and haunting in contemporary cultural theory.* Bloomsbury. [Kindle Edition]. Retrieved from amazon.ca

Eagleman, D. (2009). Metamorphosis. In *Sum: Forty tales from the afterlives*, (p. 23). Canongate.

Flyn, C. (2021). *Islands of abandonment: Nature rebounding in the post-human landscape.* Viking.

Freud, S. (1929/2018). *Civilization and its discontents.* General Press.

Government of Manitoba (2021, February) *Climate change.* https://www.gov.mb.ca/climateandgreenplan/print,climatechange.html

Harris, A., Kalb, M., & Klebanoff, S. (2016). Introduction. In A. Harris, M. Kalb & S. Klebanoff (Eds.), *Ghosts in the consulting room.* (pp. 1–16). Routledge.

Heinz, L. (2017). *Extinction: A photographic exploration by Marc Schlossman.* https://www.extinction.photo/about-the-extinction-project/

Hoye, B. (2019, March 23). 'The water passes and we're moving on'- but there are lessons in 5 of Manitoba's worst floods, expert says. CBC. https://www.cbc.ca/news/canada/manitoba/manitoba-5-worst-floods-1.5047030

Jobson, C. (2015, August 20). *Welcome to Dismaland: A first look at Banksy's new art exhibition housed in a dystopian theme park.* Colossal. https://www.thisiscolossal.com/2015/08/dismaland/

Palinkas, L. A., O'Donnell, M. L., Lau, W., & Wong, M. (2020). Strategies for delivering mental health services in response to global climate change: A narrative review. *International Journal of Environmental Research and Public Health*, 17(22), 8562. https://doi.org/10.3390/ijerph17228562

Schlossman, M. (2017a). *Extinction.* [Online exhibition of photographs]. Exhibited at https://www.extinction.photo/about-the-extinction-project/

Scholossman (2017b). *The Hawksbill Turtle.* In *Extinction.* [Online exhibition of photographs]. https://www.extinction.photo/species/hawksbill-turtle/

Sontag, S. (2007). *At the same time: Essays and speeches.* Picador.

Stibbe, A. (2021). *Ecolinguistics: Language ecology and the stories we live by*. (Second ed.). Routledge.

Stolorow, R. D. (2011). *World, affectivity, trauma: Heidegger and post-cartesian psychoanalysis*. Routledge.

Swirsky, R. (2015, November 16). Dismaland by Banksy review. Design Curial. http://www.designcurial.com/news/dismaland-by-banksy-review-4719744/

Vernick, D. (2020, July 28). *3 billion animals harmed by Australia's fires*. World Wildlife Federation. https://www.worldwildlife.org/stories/3-billion-animals-harmed-by-australia-s-fires

Winnicott, D. W. (2005). *Playing and reality*. Routledge. Originally published in 1979.

Worden, J. W. (1983). *Grief counselling and grief therapy*. Tavistock.

6 Relational and Psychoanalytically Informed Education in the Anthropocene

The epoch of the Anthropocene fuses our spiritual, cultural, political and educational prospects to the material realities of the planet. As climatological precarity escalates, humans will experience psychological jolts of increasing severity and duration. Worries about the habitability of the planet will continue to wash over the globe and the unrelenting power of unchecked capitalism will put the survival of most Beings on the planet into question. The privilege and gluttony of the very rich will be laid bare in front of the gates and security cameras they install to protect their empires. While a global system of climate apartheid expands sacrifice zones, the elite will use greater force to extract what is left of the Earth's natural resources. Affluence will buy new laws to criminalize protestors, donations will eliminate environmental regulations and campaign contributions will commodify national parks, emergency rooms, elementary schools and the water we drink. We will watch the moneyed continue to weaken the public good to protect their political, social and financial interests, unless we rupture their manufactured neoliberal illusions.

As the social contract is renegotiated by those who have a vested interest in reducing the commons, the field of education needs an ecosophical orientation informed by psychological theories to mitigate the present and future social traumas precipitated by the climate breakdown. Hoggett (2019), one of the founding members of the Climate Psychology Alliance, offers a brief description of the breadth of climate psychology:

> Climate psychology draws upon a variety of sources that have been neglected by mainstream psychology including psychoanalysis, Jungian psychology, ecopsychology, chaos and complexity theory, continental philosophy, ecolinguistics and social theory. It attempts to offer a psycho-social perspective, one that can illuminate the complex two-way interaction between the personal and the political.
>
> (p. 9)

DOI: 10.4324/9781003024873-7

I view educational encounters in the lecture hall, classroom, lab or interview room within a field of affective relations. The field "comes into being between two or more people in a way that cannot be predicted or controlled. It can only be accepted or rejected" (Stern, 2015, p. 2). My thinking about the intersubjective field is informed and inspired by the work of Donnel Stern (2003). In his first book, *Unformulated Experience: From Dissociation to Imagination in Psychoanalysis*, Stern writes about the centrality of the interpersonal field in his clinical practice:

> A fully interpersonal conception of treatment is a field theory. The psychoanalytic relationship, like any relationship, takes place in a field that is defined ceaselessly by the participants. It is not only the intrapsychic dynamics patient and analyst bring to their relationship that determine the experience with one another. The field is a unique creation, not a simple additive combination of individual dynamics; *it is ultimately the field that determines which experiences the people who are in the process of co-creating that field can have in one another's presence* [emphasis added]. It is the field that determines what will be dissociated and what will be articulated, when imagination will be possible and when the participants will be locked into stereotypic descriptions of their mutual experience. *Each time one participant changes the nature of his or her involvement in the field, the possibilities for the other person's experience change as well* [emphasis added] ...the field is the only relevant context.
>
> (p. 110)

A missing element in Stern's articulation of the field is the inclusion of the more-than-human world. As I understand the field, all Beings shape and are shaped by inter- and intra-subjective relations. Within a more inclusive description of the field of relations, the ethos of education and the obligation of the teacher shifts in significant ways. Learning is organic, multivocal and emergent and oriented towards cultivating a respect for biological and cultural diversity. Curriculum is constituted from the lived experiences of children's lives rather than from a PDF downloaded from an educational authority's website. Encounters between teachers, students and Others are explorations of the mysteries of the world and a co-venturing to discover one's place within the wider web of relations. The rigidity of rubrics gives way to dialogical encounters in which teachers respond to the unique curiosities of each child. Teachers demonstrate an attunement to students' centres of being (Benjamin, 2018). They compose themselves as compassionate witnesses and in response, students recognize themselves in the eyes of their teachers. When children feel recognized, they trust their teachers to navigate the everyday affective ruptures of social life and then count on them to facilitate relational repair when unwieldy feelings surface in classrooms. The teacher then becomes an interlocutor between the world out there and the one her students formulate in their minds (D'Amour, 2020).

Emergence is another defining quality of the intersubjective field. Psychoanalysts most commonly characterize emergence as an interaction between two people that becomes a self-organizing system. Emergence creates a third space, one containing the unique contributions of all Beings enmeshed in an encounter. To beckon emergence without being invited or demanded, a teacher orients herself in receptive attention and allows the articulation of the child's subjective experience to arrive unbidden. It requires a teacher to balance dependency with independence, particularly when a child becomes estranged as they learn new things. Working to reduce asymmetries of power in the encounter with the child, the teacher adjusts to the unpredictable variations in the child's utterances or gestures without attempting to assimilate the child's mind. Emergence in the classroom is improvisational, vibrant, and mysterious, and it can leave you sensory-struck. Situating curriculum and pedagogy within this intersubjective frame has major implications for the aims, methods and interpretations of educative experiences. Learning essentially becomes the transformation of relations between significant Others.

To make a case for relational and psychoanalytically informed pedagogy and educational research, I discuss the evolution of intersubjective psychoanalytic theory by introducing the work of Sándor Ferenczi. Building on Ferenczi's seminal work on mutual analysis, I inquire into the phenomena of transference and countertransference and how they impede mutual recognition in classrooms. I go on to expand Winnicott's concept of the holding environment by including social media's digital habitat and the more-than-human world. As I broaden the field of relations, I identify some of the dialogical and material barriers exacerbating the political conditions that are making a wide-scale response to ecological collapse exceedingly more difficult.

Sándor Ferenczi

Sándor Ferenzci (1873–1933) was one of Freud's most influential students. Repelled by the dogma of professional expertise, he surrounded himself with cultural innovators from the fields of sociology, psychology and anthropology (Haynal & Haynal, 2015). Stemming from his belief the problems his patients faced were enmeshed within the larger social context, he sought out connections with scholars and practitioners from different fields of study. Consequently, his interdisciplinary approach to psychoanalytic inquiries opened psychoanalysis to other disciplines. Ferenczi thrived within an interdisciplinary milieu, but he was also responsible for the development of organizations that focused on the dissemination of clinical and theoretical innovations in psychoanalysis.

For instance, he was a founder of the Hungarian Psychoanalytic Society (Mészáros, 2015) and he is credited for proposing the creation of the International Psychoanalytic Association. Colleagues who were closest to Ferenczi said what inspired him the most however was his work with patients who had been turned away by other analysts. To that end, he is frequently referred to as the "analyst of last resort" by other intersubjective psychoanalysts. In fact, it was a session with one of his most challenging patients Elizabeth Severn that led to several important innovations in his practice and fertilized the seeds for what we now refer to as relational psychoanalysis.

Countertransference

Holmes (2014) studied the historical legacy of the psychoanalytic concept of countertransference and determined countertransference was frequently discussed in three different ways across the literature: *interfering, useful* or *intersubjective*. Freud said little about countertransference, but when he did write about it, he described it as interfering (Berzoff & Kita, 2010; Gabbord, 2001) Alternatively, Ferenczi thought it was productive and intersubjective (Ferenczi, 1995). In a diary entry dated January 7, 1932, Ferenczi deconstructed one of his new intersubjective techniques and mused, "The patients begin to abuse my patience, they permit themselves more and more, create very embarrassing situations for us, and cause us not insignificant trouble. Only when we recognize this trend and openly admit it to the patient does this artificial obstacle, which is of our own creation disappear" (p. 2). Here we find evidence of the countertransference doing its work on him. He references "embarrassing situations" and "trouble" when encounters with his patients try his patience or trigger negative feelings. Ferenczi's work provides some preliminary groundwork to entertain moody pessimism in the classroom as well as the teachers' leaky feelings triggered by interactions with students. In many teacher education courses, there is a lot of talk about loving other people's children, but as psychoanalysis teaches us, when love is in the air, so is hate, envy and aggression. The language of countertransference is a healthy way to acknowledge and process the bad feelings that arise when a student in your class talks throughout your lecture or rolls his eyes every time you assign a reading.

In subsequent diary entries, Ferenczi refers to a patient who demands an opportunity to analyze her analyst. In these entries, you find the origins of contemporary relational psychoanalytic concepts such as mutual recognition (Abraham, 1968; Mitchell, 1988; Mitchell, 1997; Mitchell, 2000; Orange, 2011; Stern, 2005; Stern, 2015). During some of his early experiments with mutual analysis with his patient Elizabeth Severn, he described instances when he found it too difficult to listen to her. In his diary, he expressed negative feelings when she competed for his wandering attention during a session. Closer to his death, Ferenczi distanced himself from some of his more radical experiments with mutual analysis. He revealed that at times he felt out of control. Further, the emotional cost of being vulnerable in the presence of one's patients was on many occasions intolerable for him. His work is a guardrail when discussing mutuality in the context of pedagogy. Although learning encounters can trigger countertransference, teachers are obligated, professionally and morally, to intervene in unhealthy group dynamics. More importantly, the existence of countertransference does not release a teacher from the responsibility of being a mature, emotionally grounded witnessing presence in the classroom or lecture hall.

Freud was a mercurial mentor. In response to Ferenczi's experiments with mutual analysis and compounded by Freud's perception his mentee was unable to set appropriate boundaries with patients, Freud cast Ferenczi outside the inner circle. In fact, Freud went so far as to publicly refer to his former protégé as the *enfant terrible*. Thankfully, in the late twentieth century, a group of psychoanalysts re-engaged with Ferenczi's work to challenge the ambivalence towards countertransference in classical psychoanalysis (Mitchell, 1988; Mitchell, 2000; Orange, 2011). Many of the new intersubjective clinicians encouraged their colleagues to work with their own fears, desires and ambivalence when it played a part in the analytical process. The history of countertransference in the field of psychoanalysis is a way to understand what happens when the interior worlds of different individuals collide, and this understanding is integral to the work of the teacher/analyst. Ferenczi's consideration of the productive capacity of countertransference not only opened the door for psychoanalytically informed dialogues grounded in mutual recognition (Mészáros, 2015), he developed a language of intersubjectivity to describe what happens between people in the classroom or consulting room when depth and ambiguity collapse.

Relational psychoanalysis is a relatively new subfield in psychoanalysis. Greenberg and Mitchell's book *Object Relations in Psychoanalysis* in 1983 is often identified as the birth of contemporary intersubjective praxis (Greenberg & Mitchell, 1983). Benjamin, Greenberg, Mitchell and others began to frame new analytical approaches by drawing from the research on infant attachment (Bowlby, 1969; Bowlby, 1973; Stern, 1985), Sullivan's interpersonal psychology (1940; 1953; 1956) and Ferenczi's work on mutual analysis and introjection (Aron & Harris, 1993; Frankel, 1998; Frankel, 2002). Another defining characteristic of this group was their resistance to Neo-Freudian praxis that heavily relies on the assumption that self-preservation and sex are the most important drivers of human behaviour. In contrast, relational psychoanalysts believe we are born to connect with one another and that social experiences are the primary motivators of psychic life.

Mitchell (1988; 1993a; 1993b; 1997; 2000), whose work was significant in the early stages of the field's development (Ringstrom, 2010), characterized the mind as a set of relational configurations. He suggested that relational configurations influence one's predictions about the behaviour of others. They shape how one interprets other people's behaviour and they influence how one behaves with other people. Our formative relationships influence who we are and who we can become in the face of others. That said, past relationships are always with us in the present day, and often during periods of heightened emotional stress, people unconsciously enact familiar relational patterns. Echoes of relational performances from the past can make mutual recognition more difficult by limiting what can be made possible between two people in an encounter. Mitchell's insights are critical for understanding some of the emotional wildness in classrooms. When a teacher unconsciously projects his authority issues on present-day relationships, it interrupts receptive attention, offending the innocent strengthens the illusions as the means to protect his ego.

The First Holding Environment

Within a relational perspective, the environment (nursery, home, classroom, forest, beach) is part of the relational field. Young people come to know themselves by making meaning of their own reflections in the eyes of significant Others. During the first few months of their lives, baby humans navigate their intersubjective and intrapsychic worlds through what Winnicott (1979/2005) calls the first "not me" possession (p. 2) and the rhythms of affective couplings. It begins with a newborn's fist in mouth and leads to an attachment to a stuffed teddy bear or soft toy. Winnicott (1979/2005) explains:

> I have introduced the terms 'transitional objects' and 'transitional phenomena' for designation of the intermediate area of experience, between the thumb and the teddy bear, between the oral eroticism and the true object-relationship, between primary creative activity and projection of what has already been introjected, between primary unawareness of indebtedness and the acknowledgement of indebtedness (Say: "ta"). By this definition, an infant's babbling and the way in which an older child goes over a repertory of songs and tunes while preparing for sleep come with the intermediate area as transitional phenomena, along with the use made of objects that are not part of the infant's body yet are not fully recognized as belonging to external reality.
>
> (pp. 2–3)

Babies and caregivers engage by matching internal states. As little ones grow, gestures, emotional mirroring and *bubbles-of-language-on-the-way* form the basis of the intersubjective and intrapsychic lives of toddlers. Winnicott (1979/2005) asks:

> What does the baby see when he or she looks at the mother's face? I am suggesting that ordinarily, what the baby sees is himself or herself. In other words, the mother is looking at the baby and what she looks at is related to what she sees there. All this is too easily taken for granted. I am asking that this which is naturally done well by mothers who are caring for their babies shall not be taken for granted. I can make my point by going straight over to the case of the baby whose mother reflects her own mood, or worse still, the rigidity of her own defences. In such a case, what does the baby see?
> (p. 151)

Rhythmic affective fusions (and dislocations) make imprints on a child's mental topography in ways that facilitate or inhibit a young person's attachment, sense of security and their ability to tolerate the disillusion of what Winnicott (1965) refers to as the *holding environment*. The holding environment is the name he gave to describe the space in which significant adults protect the growth and development of the unique individuality of the child.

Holding refers to a protected caregiving space in which children trust their environment and feel a sense of safety. Analysts and teachers provide emotional holding by attuning to the Other's feelings without judgement. Winnicott's belief was that a child sees herself through the eyes of her mother. When mothers or other significant adults in a child's life accept the whole child and demonstrate knowledge of the child's uniqueness, the acceptance and recognition metamorphoses into a positive self-image and the capacity for self-regulation. Inevitably, and repeatedly, young children question what they hold in their minds about others. The relationship Winnicott referred to as the *good-enough-mother-child relationship* makes it possible for a child to *keep the other's mind in mind* (Fonagy & Target, 1998) and to remain emotionally intact.

Classroom as Holding Environment

Helping teachers attune to the emotional dimensions of learning and by making them aware of the phenomena of transference and countertransference can help them harmonize their responses when they are frustrated by the behaviour of students. When I was teaching grade seven English Language Arts many moons ago, one of my students, Amy, would frequently skip class, refuse to do her homework and she became oppositional when an adult asked her to do something in the school. Interactions with Amy often left me feeling frustrated, resentful and like a failure. After getting to know Amy's family over the course of the school year, I recognized Amy was enacting a similar relationship she had with her mother with me. Amy viewed her mother as pushy, demanding and judgemental. When the teaching assistant or I asked her to do something in class, Amy desperately tried to assert her autonomy by throwing her assignment in the recycling bin or by hiding the computer mice all over the classroom. Looking back, I would label her actions as transference behaviour and the increases in demands I made of her, along with my sour facial expressions, as evidence of countertransference.

A young person's security and attachment is deeply connected to the ability of a caregiver to engage in what Stern (1985) calls affect attunement. Attunement is a process in which adults communicate a reciprocal affective state without losing themselves in the process. A communion of recognition resembles elements of a Winnicottian-inspired theatre exercise I have facilitated on multiple occasions called *Mirror Instincts*. To begin the exercise, participants choose a partner and move to an open space in the room. The partners stand facing one another in close proximity. The facilitator invites the participants to allow their bodies to become a set of instinctual mirrors that will reflect each other's body movements and facial expressions. Typically, one person in each dyad takes a leadership role in the beginning of the exercise. After 30 seconds or so, I remind the players the goal of the exercise is to achieve a level of synchronicity that would make it difficult to know which person was initiating at any given time. Close observation and the willingness to remain in a state of *compassionate anticipation* is a requirement for achieving state of mutuality in the exercise. This means waiting with an open heart and trusting you and your partner are capable of recognizing and being recognized.

After the exercise is over, I ask the participants what they noticed. They usually make comments on the impermanence of synchronicity and lament how difficult it is to stay attuned to the other. When facilitating this exercise with teacher candidates and more experienced teachers, I invite them to think metaphorically about Mirror Instincts and relationality in the classroom. On several occasions, I have been sensory-struck by a connection a teacher has made between the game and what it means to take responsibility for the Other within educative encounters. My experience teaches me it is often affective cues that move participants closer to a mirror image. On one occasion, I saw a participant wince when she bent her knee during the game. Her partner winced too. As they both extended their bodies upward to distance themselves from the pain of the knee bend, a silent agreement was made to conduct the rest of the exercise using the top half of their bodies. As the exercise progressed, I noticed their arm and hand movements became more nuanced and their faces expressed a broader range of emotions. In the last part of the exercise, from across the room, they appeared frozen in time. As I approached them, I realized they had located the exercise in their eyes.

For the next two minutes, they managed to communicate joy, wonder, fear and sarcasm through eye rolls, winking and wide-eyed gazes.

Mirror Instincts requires the participants to remain open to the unbidden, similar to the way teachers move about their students' mental topographies before speech is palpable. Yes, teachers need mirror instincts, but if intersubjectivity is to be realized in the classroom, the student must be involved in creating and maintaining the exchange of recognition. We need both actors to play Mirror Instincts. Sometimes students describe what I would label as the emergence of the uncanny during the exercise, because they cannot rationally explain why they are capable of mirroring the Other with such nuance and compassionate anticipation. When the uncanny surfaces in the conversation, it opens the opportunity to address the collective unconscious. I describe the collective unconscious as a part of the mind filled with ancestral sensory memories, the universal recognition of concepts and an instinctual recognition of archetypal relationships (mother/child, teacher/student, analyst/analysand) in Others. Relational psychoanalysis pulls our attention to the enigmatic unconscious.

The Earth as the Holding Environment of All Holding Environments

If the Earth is the holding environment of all holding environments, we are deeply disoriented right now. As species disappear, oceans acidify and forests are clear-cut, melancholy, exasperation, and fear overwhelm numbness and outrage bubbles up from the unconscious. Instead of doing the necessary work on the foundations of our shared holding environment, adults expect young people to show up at school and act as if everything is normal (Hickman, 2019). To evade the unmentalized fear that comes from knowing too much about the danger climate changes pose, much of their collective angst seeps into nightmares and the murky territory of the collective unconscious.

Orange (2017, g. 39) in her book *Climate crisis, Psychoanalysis, and Radical Ethics* quotes Hans Loewald's (1960) description of the unconscious as a shadow-life of climate change:

> The transference neurosis, in the technical sense of the establishment and resolution of it in the analytic process, is due to the blood of recognition which the patient's unconscious is given to taste-so that old ghosts may reawaken to life. Those who know ghosts tell us they long to be released from their ghost-life and led to rest as ancestors. As ancestors they live forth in the present generation, while as ghosts they are compelled to haunt the present generation with their shadow-life. Transference is pathological in so far as the unconscious is a crowd of ghosts, and this is the beginning of the

transference neurosis in analysis: ghosts of the unconscious, imprisoned by defenses but haunting the patient in the dark of his defenses and symptoms, are allowed to taste blood, are let loose.

(p. 29)

Orange goes on to discuss the "unburied crimes" and the "collective ghosts" of settler colonialism and slavery that continue to haunt the relational field. In Canada, the plethora of water advisories in First Nations communities are examples of unburied crimes. When teachers do not acknowledge these intergenerational injustices in their classrooms, it can profoundly affect their students' sense of agency, moral imagination and capacity to mentalize because the dismissal causes young people to incorporate instead of introjecting the painful emotions discharged from unburied ecological crimes.

Young children eventually become aware drilling for oil and gas damages the planet, but they also appreciate getting a ride to soccer practice in their parent's pick-up truck. They are told to be kind to animals and then asked to flip the steaks on the barbecue grill for dad. Recycle your pop can and then get on the plane to Disneyworld. Ambivalence is fantasizing while you are spinning in place. I can relate. After each report issued by the IPCC, I often fantasize the Task Force on National Greenhouse Gas Inventories (TFI), the group responsible for developing and refining an internationally agreed upon methodology for the calculation of greenhouse gas emissions, experienced a glitch in the matrix. In my fantasy, a very serious looking person goes on a mea culpa tour to tell everyone the good news; the TFI bought software on sale and the programmers who forecasted global warming is likely to reach 1.5 degrees Celsius above pre-industrial levels between 2030 and 2052 were wrong. Un/fortunately, another report will cross my desk and snap me out of my soothing illusion. I go back to holding the unmentionable while I fly sheepishly to give a 15-minute talk at an academic conference.

Anti-Social Media as a Leaky Holding Environment

The climate emergency has generated a psychosocial rupture in life's continuity. As we question the safety of our environment and, in turn, our trust in Others, the leaders of big tech companies responded by further exploiting our anxiety. By weaponizing resentment and fear generated by nefarious social constructions of race, sexuality, gender, language and religion, Silicon Valley has exacerbated the political conditions as to make a wide-scale response to ecological collapse exceedingly more difficult. One only has to examine the responses to social media posts from leaders like Borris Johnson or Donald Trump to see how the slogans "Make America Great Again" and "There are no disasters, only opportunities" have wreaked havoc on civil discourse. Scapegoating immigrants and feasting on white grievance, these tactics and the tactics of other strongmen leaders are part of a larger historical pattern of powerful white men degrading relationality by manipulating public discourse to prevent social justice movements from consolidating and gaining momentum. Now, Facebook, Twitter, Youtube and Instagram allow these kinds of characters and their supporters to tweet fractious slogans 24 hours a day.

Relational psychoanalysis teaches us that nothing exists in isolation. We come to know ourselves and our environment within a web of interconnections. In fact, my ability to function as a healthy person in the world is dependent on my capacity to trust significant others in my environment. Psychologists refer to this ability as *epistemic trust*. Epistemic trust describes a person's ability to predict new information from another person is authentic, generalizable and relevant to oneself (Fonagy & Allison, 2014, p. 373). Epistemic trust not only shapes our expectations and interpretations of other people's words and actions, it also informs our assessments of the interior worlds of others, what Fonagy et al. (2002) calls *metalized affectivity*. One of the most destructive consequences of the clickbait economy is the corruption of mentalization.

When we mentalize, we base our response to another person's behaviour on conscious and unconscious perceptions of someone's intentions. There are two types of metallization. *Explicit mentalization* is the process of turning feelings into words. It is a conscious and reflexive process, whereby we narrate our experiences of the world (Knox, 2016). Explicit mentalization is the ability to know something true about the world and then to symbolically express one's knowing to Others. However, most mentalization is implicit and unconscious (Van Overwalle & Vandekerckhove, 2013). *Implicit mentalization* is rooted in the affective dimensions of communication and it emerges from the patterns of relations we inherit in the context of our significant relationships. For instance, if your father was often critical of you in public, you might assume negative intentions of those who respond critically to an article you posted on Facebook or you might spend too many hours coveting likes on Twitter to fill a void left by the absence of your father's validation.

Over the last 10 years, increasing psychological distress and treatment for mental health among youth in North America has paralleled the sharp climb in their use of social media (Abi-Jaoude et al., 2020; Karim et al., 2020). "Evidence from a variety of cross-sectional, longitudinal and empirical studies implicate smartphone and social media use in the increase of mental distress, self-injurious behaviour and suicidality among youth; there is a dose response relationship, and the effects appear to be greatest among girls" (Abi-Jaoude et al., 2020, pg. E136). In the siloed social media habitus constructed by big tech, emotional contagion makes it harder for young people to trust the minds of others, especially when the Other is perceived not to look, act, speak, worship or vote like oneself. It is hard to be your true self and to believe that other people are being authentic within the *likes commodity exchange.*

In the 'real' world, young people's epistemic trust and faith in significant adults is being eroded as the Earth's habitability decreases and the uncertainty of a livable future increases. In their digital world, disaster stressors such as news reports of climate volatility, amplify feelings of exclusion, stress, loneliness and anxiety in an environment that makes it very difficult to assess the interior worlds of others. In fact, the digital world's clickbait economy profits from the widespread corruption of mentalized affectivity by rewarding allegiance to social silos and inhibiting the consideration counternarratives through clever algorithms. In the trailer for the documentary *The Social Dilemma*, the narrator warns when a person types "*climate change is...* into Google, you are going to see different results depending on where you live and the particular things Google knows about your interests. That's not by accident, that's a design technique" (Orlowski, 2020). Part of cultivating solidarity in the digital age, then, requires us to reclaim our collective agency by unplugging. Freed from Google's data harvesting, in the material world of schools, teachers and students could research, for example, the percentage of their carbon fair share, and in the method and manner of the inquiry sow some epistemic trust in mutually recognizing and responding to the calculations. Philip Cushman (2012) writes about the political implications of working relationally:

> Relationalists live out certain moral understandings that make political commitment possible: in their work they value honesty, fairness, collaboration, egalitarianism, freedom. Their commitments to compassion, self-reflection, and intellectual integrity make their analytic practices one of the few places where a resistance against the anti-intellectualism, militarism, racism, misogyny, classism, and homophobia of the far right come to light. It is that realization that has moved me to interpret relational psychoanalysis as a kind of quiet political resistance.
>
> (Cushman, 2012, pg. 12)

Relational approaches to the study and practice of education run counter to the hyper-individualistic, quick fix culture permeating the field. Ecosophical education and research actively challenges neoliberal entitlements such as: (a) demanding recognition while expecting others to go without, (b) being spared from messy feelings, (c) building escape routes to evade one's obligations to Others, (d) living beyond the Earth's biological limits and (e) calculating one's worth by adding up the "likes" harvested from social media. To problematize, and then work against the entitlements, we must identify what is eroding the social containing environment (Weintrobe, 2021) and then educate and research in support of a culture of care in the digital and in the more-than-human world.

Relationship talk in the field of education often overemphasizes what adults should say or do for children at the expense of considering the relational potency and potential of making space for the child to initiate a communicative act. But it is a mistake to view emergence as one-way-transmission. For a child to evolve and mature, he needs a significant Other to recognize his unique communicative offerings. Benjamin (2018) uses the term *rhythmic thirdness* to describe the continuous mutual adjustment that allows the teacher/student to acknowledge their differences while striving for mutual recognition. When teachers engage in this complex dialogical space, artfully navigating the tension between dependency and independence, they create and maintain strong social bonds. Most importantly, when a child is provided a space to recognize her teacher's individuality without incorporating the teacher as the bad object in the learning space, the child is given a chance to remake her identity and to form healthy attachments in the eyes of a significant Other.

Education both in its untimely focus on individual attainment and social mobility is out of time. What is called for now are relational pedagogies and research methodologies to adapt to the affective ecologies entangling one's dialogic self in the Anthropocene. Valuing an exploration of one's inner lifeworld in relation to Others is a place to begin, so love, care, compassion and forgiveness are brought to bear in complex conversations about what it means to become an educated person in the Anthropocene.

References

Abi-Jaoude, E. N., Treurnicht, K., & Pignatiello, A. (2020). Smartphones, social media use and youth mental health. *Canadian Medical Association Journal (CMAJ)*, *192*(6), E136–E141.

Abraham, N. (1968). The shell and the kernel: The scope and originality of freudian psychoanalysis. In Abraham, N., & Torok, M. (1994). *The shell and the kernel: Renewals of psychoanalysis*. (pp. 79–106). (Vol. 1). (N. T. Rand, Trans.). University of Chicago.

Aron, L., & Harris, A. (Eds.) (1993). *The legacy of sándor ferenczi*.The Analytic Press.

Benjamin, J. (2018). *Beyond doer and done to: Recognition theory, intersubjectivity and the third*. Routledge.

Berzoff, J., & Kita, E. (2010). Compassion fatigue and countertransference: Two different concepts. *Clinical Social Work Journal*, *38*, 341–349. doi: 10.1007/s10615-010-0271-8

Bowlby, J. (1969). *Attachment*. Basic Books.

Bowlby, J. (1973). *Separation: Anxiety and anger*. Basic Books.
Cushman, P. (2012). Book review: Why resist? Politics, psychoanalysis, and the interpretive turn. *DIVISION: A Quarterly Psychoanalytic Forum, 4*(3), 11–13.
D'Amour, L. (2020). *Relational psychoanalysis at the heart of teaching and learning: How and why it matters*. Routledge.
Ferenczi, S. (1995). *The clinical diary of sándor ferenczi*. Judith Dupont (Ed.). Translated by M. Balint & N. Z. Jackson. Harvard University Press.
Fonagy, P., & Allison, E. (2014). The role of mentalizing and epistemic trust in the therapeutic relationship. *Psychotherapy, 51*(3), 372–380. doi: 10.1037/a0036505
Fonagy, P., Gergely, G., Jurist, E., & Target, M. (2002). *Affect regulation, mentalization and the development of the self*. Other Press.
Fonagy, P., & Target, M. (1998). Mentalization and the changing aims of child psychoanalysis. *Psychoanalytic Dialogues, 8*(1), 87–114.
Frankel, J. (1998). Ferenczi's trauma theory. *American Journal of Psychoanalysis, 58*, 41–61.
Frankel, J. (2002). Exploring Ferenczi's concept of identification with the aggressor: Its role in trauma, everyday life, and the therapeutic relationship. *Psychoanalytic Dialogues, 12*(1), 101–140.
Gabbord, G. O. (2001). A contemporary psychoanalytic model of countertransference. *In Session: Psychotherapy in Practice, 57*(8), 983–991.
Greenberg, J. R., & Mitchell, S. A. (1983). *Object relations in psychoanalytic theory*. Harvard University Press. Kindle Edition.
Haynal, A. E., & Haynal, V. D. (2015). Ferenczi's attitude. In S. K. Wang, A. Harris & S. Kuchack (Eds.), *The legacy of sándor ferenczi: From ghost to ancestor*. (pp. 52–74). Routledge.
Hickman, C. (2019). Children and climate change: Exploring children's feelings about climate change using free association narrative interview methodology. In P. Hoggett (Ed.), *Climate psychology: On indifference and disaster*. (pp. 41–59). Palgrave Macmillan.
Hoggett, P. (2019). Introduction. In P. Hoggett (Ed.), *Climate psychology: On indifference and disaster*. (pp. 1–19). Palgrave.
Holmes, J. (2014). Counter-transference in qualitative research: A critical appraisal. *Qualitative Research, 14*(2), 166–183. 10.1177/1468794112468473
Karim, F., Oyewande, A., Abdalla, L. F., Chaudhry Ehsanullah, R., & Khan, S. (2020). Social media use and its connection to mental health: A systematic review. *Cureus (Palo Alto, CA), 12*(6), e8627–e8627. https://doi.org/10.7759/cureus.8627
Knox, J. (2016). Epistemic mistrust: A crucial aspect of mentalization in people with a history of abuse? *British Journal of Psychotherapy, 32*(2), 226–236. doi: 10.1111/bjp.12212
Mészáros, J. (2015). Ferenczi in our contemporary world. In A. Harris & S. Kuchack (Eds.), *The legacy of sándor ferenczi: From ghost to ancestor*. (pp. 19–32). Routledge.
Mitchell, S. A. (1988). *Relational concepts in psychoanalysis: An integration*. Harvard University Press.
Mitchell, S. A. (1993a). *Hope and dread in psychoanalysis*. Basic Books.
Mitchell, S. A. (1993b). Aggression and the endangered self. *Psychoanalytic Quarterly, 62*, 351–382.

Mitchell, S. A. (1997). *Influence and autonomy in psychoanalysis*. The Analytic Press.

Mitchell, S. A. (2000). *Relationality: From attachment to intersubjectivity*. The Analytic Press.

Orange, D. (2011). *The suffering stranger: Hermeneutics for everyday clinical practice*. Routledge.

Orange, D. (2017). *Climate crisis, psychoanalysis, and radical ethics*. Routledge.

Orlowski, J. (Director). (2020). *The social dilemma [documentary film]*. Netflix.

Ringstrom, P. A. (2010). Meeting Mitchell's challenge: A comparison of relational psychoanalysis and intersubjective systems theory. *Psychoanalytic Dialogues, 20*, 196–218. doi: 10.1080/10481881003716289

Stern, D. (1985). *The interpersonal world of the infant*. Basic Books.

Stern, D. (2005). Intersubjectivity. In E. S. Person, A. M. Cooper, & G. O. Gabbard (Eds.), *Textbook of psychoanalysis* (pp. 77–92). American Psychiatric Publishing.

Stern, D. B. (2003). *Unformulated experience: From dissociation to imagination in psychoanalysis*. Routledge.

Stern, D. B. (2015). *Relational freedom: Emergent properties of the interpersonal field*. Routledge.

Sullivan, H. S. (1940). *Conceptions of modern psychiatry*. Norton.

Sullivan, H. S. (1953). *The interpersonal theory of psychiatry*. Norton.

Sullivan, H. S. (1956). *Clinical studies in psychiatry*. Norton.

Van Overwalle, F., & Vandekerckhove, M. (2013). Implicit and explicit social mentalizing: Dual processes driven by a shared neural network. *Frontiers in Human Neuroscience, 7*, article 5600. doi: https://doi.org/10.3389/fnhum.2013.00560

Weintrobe, S. (2021). *Psychological roots of the climate crisis: Neoliberal exceptionalism and the culture of uncare*. Bloomsbury.

Winnicott, D. W. (1965). *The maturational process and the facilitating environment*. International Universities Press.

Winnicott, D. W. (2005). *Playing and reality*. Routledge. Originally published in 1979.

7 Cultivating Solidarity as the Climate Crisis Intensifies

In 1849, Eunice Foote, one of the first climate scientists, made a remarkable discovery; when sunlight shone on carbon dioxide (CO_2) in a closed container, heat built up inside. On August 23, 1856, a reading of Foote's work was shared at the American Association for the Advancement of Science. In her paper, Foote suggests that an atmosphere of CO_2 would increase the earth's temperature (Foote, 1856, p. 382). Soon after, her discovery was attributed to Irish physicist, John Tyndall, and once he received the credit, her name faded from the origin stories of climate science. The rest is his/story.

The disappearance of Eunice Foote from science and history textbooks illuminates one of the barriers to cultivating solidarity amid the climate crisis. Foote's work, like so many other women scientists, was ignored, disappeared or attributed to someone else. Rossiter (1993), refers to the phenomenon as *The Matilda Effect*, named after Matilda Joslyn Gage (1826–1898), who in 1893 published *Woman, Christ and the State*, "…to show how Christianity has justified and extolled the subjection of women – it urges them to work hard and to sacrifice, it takes their money, but in return it gives them little credit and even justifies men's exploitation of them" (Rossiter, 1993, p. 336). She also worked with a group of 20 other women to write the Women's Bible to reinterpret the Bible within a feminist framework. Rossiter laments that Gage remains a cipher today. She goes on to argue that "Matilda" is a fitting symbol to represent the many women in science who have had their work maligned, misappropriated or missed.

The climate amplifies and intensifies how those with privilege (gender, race, class, sexual orientation) collude in the maintenance of systems of oppression. Black, Indigenous and people of colour (BIPOC) are disproportionately impacted by the detrimental effects of the climate emergency, yet their voices are underrepresented within the field of ecology (Duc Bo Massey et al., 2021). The overwhelming whiteness in scientific fields such as ecology and evolutionary biology impacts research agendas, grants, mentorship opportunities, peer review processes and the work of faculty search committees. Facing theses systemic challenges and others,

DOI: 10.4324/9781003024873-8

BIPOC scholars face many systemic barriers related to recruitment and retention in their respective fields in science. The roots of the ecological crisis grow from these systems of oppression. Sexism, racism, heteronormativity, colonization, anthropocentrism and the injustices produced in the intersections of these systems penetrate the collective psyche and obfuscate the disproportionate effects of environmental crises and climate precarity on BIPOC communities.

Levinas (1969; 1998) reminds us that the forgetting of self moves justice. Only a radical commitment to solidarity will generate the kind of robust and inclusive responses needed to mitigate the damage wrought by the extinctions of different species, monocultures, gigafires, floods, droughts, sea level rises, severe weather events, food scarcity and pandemics. Part of the forgetting of self means one's obligation to the Other is present even before he, she or they arrives. The climate crisis grows from the exploitation of people and the more-than-human world in sacrifice zones. To turn away from one's complicity in the slow violence of climate change is a moral obscenity. Whereas Levinas understands the turn to the Other as a moral imperative, his call runs counter to the values of individualism and the valorisation of independence that form the foundation of colonial enactments of education. Keeping the disproportionate impacts of the climate crisis on BIPOC students and their families in mind, in this chapter I respond to Levinas's call and make the case that education at this time in our lives should be an answer to the question, *how can I become my Other's keeper?*

I chose to use the word *Other* instead of the word *sister* or *brother* in this chapter's guiding question not only to put some distance between Cain and Abel and myself, but because a defining impediment to cultivating solidarity is anthropocentric hubris. In the field of education, humans have constructed an *ouroboros-subjectivity* in which they determine their relative importance to other Beings in relation to themselves. Swallowing our own tails, we created fantasies of omnipotence, grandiose short-term thinking and destructive economic myths that have led to the climate emergency. How we narrate our lives matters. If we want to avoid an apocalyptic ending for the human species, we need to tell new stories about the relationships among humans and their more-than-human relatives.

Ecolinguistics is a budding area of study that invites the critical examination of what Stibbe (2021) calls "the stories we live by" (p. 2). It is a frame that can be used to deconstruct the forms of language that contribute to ecological degradation (p. 1). Rethinking terms and their legacies can lead to deeper dialogue about plants in city spaces that move us beyond *plants-as-building décor* and towards deeper dialogue about urban biodiversity. As another example, we could change the language of ownership used to refer to family pets. Instead of *dog owner* we could use the term *family member*. Extending the invitation further from the place we rest our heads, we might resist describing Others in the world as *natural resources* or *property*. Instead, we might refer to other Beings as *persons, neighbours, cohabitants* or *mitákaye oyás' in*, which means *all are related* in the Lakota language (Lapointe, 2020).

To investigate the power structures in the stories that form the foundation of our unsustainable civilization, Stibbe (2021) not only refers to traditional narratives but also to the discourses and metaphors that frame a particular worldview. Two of the most dangerous discourses imbued in our ouroboros-subjectivity in education are that humans live outside of nature and that we are the most important Beings on Earth. As an illustration, these two stories create the conditions in which people can be ambivalent about eating meat. Raising animals for food requires large swaths of energy, land and water. Livestock production accounts for 70% of agricultural land use and is responsible for 18% of greenhouse gases such as nitrous oxide and methane (David Suzuki Foundation, 2020). Modern agricultural practices violate many animal interests such as an animal's ability to live in natural conditions, to enjoy the normal social life of its species and to live free from pain. Animals are sentient Beings with complex nervous systems who feel pain. If humans believe that other sentient Beings have rights, killing them for food is morally wrong. Most importantly, the most elemental right of any Being is to be treated as an end in oneself. Eating animals as a means for human consumption turns the lives of animals into a means for human ends. The ambivalence people feel while eating meat is not only a story of Anthropocentrism, it is also a story of the disavowal of one's complicity in the anthropogenic causes of climate change.

A landmark report (IPBES, 2019) from the Intergovernmental Science Policy Platform on Biodiversity and Ecosystem Services (IPBES) estimates the current rate of global species extinction is tens to hundreds of times higher compared to the last 10 million years and the rate is accelerating. Other key findings from the report include:

- There has been a 30% reduction in global terrestrial habitat integrity caused by habitat loss and deterioration.
- There has been a 20% decline in average abundance of native species in most major terrestrial biomes, mostly since 1900.
- Up to one million species are threatened with extinction, many within decades.
- Almost 33% of reef forming corals, sharks and shark relatives and 33% of marine mammals are threatened with extinction.
- Plus/minus 560 (+/− 10%) domesticated breeds of animals were extinct by 2016.
- The distribution of 47% of terrestrial flightless mammals and 23% of threatened birds may have been negatively impacted by climate change.

We know there have been five mass extinctions over the last 450 million years when the diversity of life on the planet has dramatically decreased. Each of the previous mass extinctions was caused by "catastrophic alterations to the environment (Ceballos et al., 2020, p. 13596). You and I are living in the sixth mass extinction. What makes this mass extinction different, relative to the first five, is the disappearance of species is accelerating at an alarming pace. Over 400 vertebrate species became extinct in the last century that would have "taken up to 10,000 years in the normal course of evolution" (Ceballos et al., 2020, p. 13597) making one of the psychological conditions of the Anthropocene the widespread disavowal of the disappearance of thousands of Beings right before our eyes (Kolbert, 2014). "Millions of populations have vanished in the last 100 years, with most people unaware of their loss" (Ceballos et al., 2020, pg. 13597). What do we make of the moral compasses of human beings when they turn away from the disappearance of "millions of populations"?

Even when there is a conscious and concerted effort to mitigate species loss, humans impose a hierarchy of importance. Most of the attention about critically endangered species is given to large and majestic looking endangered animals. For example, the critically endangered Bornean Orangutan of eastern Kalimantan and the Black Rhino of Namibia (World Wildlife Federation, 2020) are covered more extensively in visual media awareness campaigns (Goulson, 2020). In contrast, the endangered olive green metallic Salt Creek Beetle of Nebraska receives far less attention and concern from the wider public. For this reason and others, I begin this section by discussing the significance of the tiniest Beings on the planet in relation to cultivating solidarity amid the climate breakdown.

The bulk of animal life on Earth is comprised of invertebrates like insects. There is estimated to be four million species of insects that we have left to discover (Stork et al., 2015). Although the number of insects on Earth far outweighs the number of vertebrates, the impacts of the climate emergency on insect populations is less understood relative to the plight of endangered vertebrates. Recently, there has been an uptick in the number of studies published on the accelerated decline of insects across the globe. In one scientific review of all existing evidence (73 studies) on insect declines, researchers found that the local extinction of insect species was eight times faster than that of vertebrates (Sanchez-Bayo & Wyckhuys, 2019), and 41% of insect species are currently threatened with extinction (Goulson, 2020, p. 4). The most widely publicized studies on insect declines was released by the Krefeld Society, a group of entomologists who have trapped flying insects in malaise traps on 63 nature reserves across Germany for many years (Hallmann et al., 2017). They found the overall biomass of insects caught in their traps declined by 75% in the period between 1989 and 2014.

An endangered invertebrate experiencing one of the most dramatic declines in the last two decades is the monarch butterfly. Tattooed in two places on my body, I have tried to commemorate their beauty, acknowledge their precarity and mourn their losses. Biologists and backyard gardeners refer to the disappearance of these butterflies as a "silent spring" (Agrawal, 2019). Although it is difficult to determine with certainty why monarch populations are plummeting, genetically modified crops, habitat degradation, land use changes (LUCs), pollution and insecticides are the factors scientists are investigating to find out why, for example, the population of monarch butterflies fell by 90% in the last 20 years in the US. There is growing consensus among scientists that changes in the growth and flowering of different milkweed plants in North America is part of the problem. Milkweed is the host plant monarchs use as a food source in their larval form, and recently, milkweeds are not predictably flowering during the monarch migration period. Boyle et al. (2019) identified a persistent decline in eight milkweed species over several decades. In particular, the abundance of the non-weedy milkweed *Asclepias*, with over 100 species in North America, appears to be threatened.

Unlike the monarch butterfly, the magnitude of the losses of lesser known insects is dizzying. Entomologists suggest that one reason the disappearance of insect populations remains underreported is because insects are hard to track. When such losses do reach public awareness, it is often through informal reporting by citizen-entomologists, people who may notice fewer insects along the park paths they walk along during the summer months or when neighbours informally discuss the absence of fluttering wings around their porch lights at night. In fact, entomologists have coined a term for the sudden awareness of the absence of a particular group of insects: "the windshield phenomenon" (Jarvis, 2018, para. 7). The disappearance of populations of insects, what some media outlets now refer to as the "insect apocalypse" (Hunt, 2019; Resnick, 2019), will continue to have profound impacts on the lives of humans. Although long-term studies of insect diversity trends are rare (Homberg et al., 2019), one thing that biologists do agree on is that the precipitous decline in terrestrial insect species richness and declining populations is cause for concern (Habel et al., 2019). Insects are incredible recyclers of nutrients, they are food for thousands of larger animals like birds and fish and they perform essential tasks like pest control and crop pollination which make them essential to the functioning of all ecosystems. It is estimated that three quarters of all the crops grown by humans require pollination by insects (Goulson, 2020, p. 5). In fact, the whole human agricultural infrastructure depends on at-risk populations of insects to control the populations of insects who will compete for the crops we grow to put food on the table. To put it starkly, if certain populations of insects disappear, humans will too.

The disappearance of insects across the globe highlights four of the foundational problems that characterize the stories we live by in Anthropocene. First, we tend to make the climate crisis a minor character even when it plays a much larger role in the stories we tell about ecological degradation. For instance, many of the recent studies on insect population declines pull our attention to pesticide use and industrial-scale intensive agriculture, but they often minimize the role global warming plays in the demise of terrestrial insect diversity and populations. For example, entomologists tell us there is much to learn about the effects of droughts, changing precipitation patterns, atmospheric nitrification related to our dependency on burning fossil fuels (Wagner, 2020) and insect extinction.

Second, our stories of faith in real and imagined *saviour technologies* are laid bare in relation to the demise of insects. The technology (machine and seed), used in industrialized farming operations in the name of producing large quantities of cheap food for the masses, is done at a significant cost to the Earth's ecosystems. Lobbyists and shareholders promise a *Green Revolution* to end world hunger, even though they are aware the industry's practices hurtle us closer to widespread food insecurity and climate catastrophe. Rotting at the bottom of the grain bin are stories of dramatically diminishing species diversity that come from monocultural food production systems. Corporate interests collude with coveted delusions that human ingenuity will somehow outperform the intricate evolutionary systems Mother Nature has perfected over thousands and thousands of years. This fetishization of technology has become a permission slip to live out of balance, a psychosocial tool to keep eco-anxiety at bay and a justification for giving up our stewardship to those who would leave no asset stranded.

In the documentary *David Attenborough: A Life on Our Planet* (2020), Attenborough describes how the relative regularity of seasonal temperatures that made farming possible in the Holocene (Fothergill et al., 2020) allowed humans to farm amok by repurposing land to accommodate large industrial farming operations. For centuries, farmers used to save a portion of their harvest to select seeds to replant in the spring. Seed engineers and their corporate benefactors saw an interruption of the common practice of seed saving as a chance to corner and commodify the seed market. The creation and patenting of hybrid seeds has become one of the most perverse abdications of our collective responsibility for biodiversity. Corporations like Monsanto-Bayer and Dupont use lobbying and lawsuits to fiercely enforce patents of genetically modified seeds.

Infiltrating food chains all over the world, Monsanto uses a number of hardball tactics (sending private investigators to surveil small farmers, suing small farmers) to protect its genetically modified seed patents (Barlett & Steele, 2008; Totenberg, 2013). In 1998, Percy Schmeiser, a farmer from Saskatchewan, Canada, made headlines when he sued the multibillion dollar bio-tech mammoth. Schmeiser argued Monsanto's seeds landed on his farm accidentally and that he was the rightful owner of the seeds after they landed on his physical property. Monsanto's fleet of lawyers argued Schmeiser infringed on the company's patent rights when Monsanto's genetically modified Canola seeds were found on Schmeiser's property. They argued when farmers use what the company refers to as "Roundup Ready Canola', they have to purchase new seeds every year. After the Supreme Court of Canada found in favour of the corporation, Trish Jordan, Monsanto's Public and Industry Affairs Director, declared, "It has been nor will it ever be our policy to exercise our patent rights where Roundup-tolerant crops are present in a farmer's field as a result of an inadvertent or unexpected act" (Salvian, 2018, para. 10).

Monsanto's domination and control of the cotton seed sector has played out with tragic consequences in India. In 1988, the Seed Policy imposed by the World Bank required the Government of India to deregulate the seed sector. The decision made it possible for Monsanto to gain control of 95% of the cotton seed market. Debt traps, hopelessness, suffering and death have followed. Vandana Shiva, the Indian activist and founder of the Research Foundation for Science, Technology and Ecology, maintains that Monsanto's monopoly of seeds is directly related to the sharp increases in farmers' suicides in India. In an opinion article for *Aljazeera*, published in 2013 (Shiva, 2013), she stated:

> Monsanto and its PR men are trying desperately to delink the epidemic of farmers suicides in India from its growing control over the cotton seed supply. For us it is the control over seed, the first link in the food chain, the source of life which is our biggest concern. When a corporation controls seeds, it controls life. Including the life of our farmers.
>
> (para. 1)

Growing from the underbelly of Monsanto's promises of greater yields and pest control promises are superweeds and *superinsects*. Unsurprisingly, some insects have developed a resistance to Monsanto's insecticide-infused seeds. In 2011, *The Wall Street Journal* reported the entomologist Aaron Gassmann's finding that western corn rootworms in four Iowa farm fields were able to resist the built-in pesticide in Monsanto's corn plant. Reports of the rootworms' resistance caused some Iowa farmers in the northeastern region of the state to buy seeds from one of Monsanto's competitors and to go back to using harsher synthetic insecticides (Kilman, 2001). Even Monsanto acknowledged back in 2010 that in industrial farming regions in India, a cotton-attacking insect called the bollworm had developed resistance to their dominant

Bt cotton plant (Philpott, 2011). Since the commercial introduction of GMOs in 1996, Monsanto (now referred to as Monsanto-Bayer after the massive German chemical company bought it in 2018) has developed and enacted a complex and effective system of patenting, royalty collection and surveillance (Peschard & Randeria, 2020) to protect their financial interests. Although Monsato-Bayer suffered some legal setbacks in the US between 2018 and 2019 due to the toxicity of its herbicide Roundup, this period was an anomaly in the corporation's aggressive litigious history. In fact, prior to 2018, Monsanto had won every lawsuit it filed in Canada and the US since 1997 (Schapiro, 2018).

More recently, the Delhi High Court in India issued a landmark decision in the patent infringement case brought by Monsanto against its sublicensee, Nuziveedu. Correspondingly, in May 2019, the Competition Commission of India issued a report stating Monsanto had "abused its dominant position in India by charging unfair trait fees and criticized several of its practices, such as insisting on the use of proprietary hybrids, charging different prices in different regions and not abiding by government set prices" (Peschard & Randeria, 2020, p. 802). The court cases in Canada, the US and India highlight the pernicious regulatory hijacking by corporate actors at the expense of all other Beings on the planet. Furthermore, Monsanto-Bayer's corporate narratives highlight the need for counternarratives that amplify public doubt about saviour technologies that promise benefits, but in reality change what should be public into private assets for a few wealthy actors.

A minimization of the disappearance of insect populations and the related consequences to human food systems defies logic. This third story we live by, that humans can live outside the ecosystems they inhabit, may expedite the extinction of humans. To put it in concrete terms, the system that currently puts food on many tables across the world will lead to increased food shortages and widespread starvation because industrial farming is pushing us into direct competition with insects for the crops that other endangered insects once helped to pollinate. Robert Jay Lifton (2017) refers to a growing awareness of this kind of absurdity as the *climate swerve*. He takes the word *swerve* from the work of Lucretius, the Roman philosopher who lived in the first century BCE. Lucretius used the term *swerve* to describe the unexpected movement of the small particles he believed constituted our universe (p. 101). Lifton notes that as we become more conscious of the climate crisis, "our efforts at adaptation have included reckless consumption of the planet's energy resources notably its fossil fuels, while numbing ourselves to the recognized consequences" (p. 4). Our ambivalence about our fossil fuel dependency, coupled with young people's demands that adults take responsibility for the eco-crisis, elicits a pervasive psychosocial tension across society. And it is within this oscillating epicentre of doubt, terror and awareness that the power of the swerve grows.

The productive discomfort can be harnessed to focus our attention in new directions, meaning the existential anxiety that comes from looming fires, floods and food shortages can be a space to generate the energy needed to act. It must be said that many in the Global South are already experiencing the psychological and material effects of the climate emergency. That said, I believe that we can use the existential anxiety that is beginning to permeate adult minds in the Global North to move towards a different, more gentle and compassionate future for all Beings on the planet:

> By confronting dire catastrophe and taking in the resulting death anxiety, even the possible death of our species, we make the swerve possible. And the swerve itself has an integrative effect that can, in turn, utilize the increasingly conscious death anxiety. That death anxiety, no longer avoided, becomes a stimulus for a continuous dynamic of awareness and potential action. In that way, the swerve creates a state of mind appropriate to the threat. And death anxiety becomes an animating force that both enhances, and is kept in check by, the swerve.
>
> (p. 104)

Anthropocentrism frames the fourth and one of the most damaging stories we live by in the fields of psychoanalysis and education. In the early days of psychoanalysis, the rationalizations used to deny other Beings the same dignity and respect that humans claim for themselves was enacted through a language of dominance, one that denied animals' capacity for reason and the possession of a soul:

> Not content with this supremacy, however, [man] began to place a gulf between his [sic] nature and theirs. He denied the possession of reason to them, and to himself he attributed an immortal soul, and made claims to a divine decent which permitted him to annihilate the bond of community between him and the animal kingdom.
> (Freud, 1917, p. 140)

"Through the oedipal production of shame" (Gentile, 2018, pg. 8), Freud carved anthropocentrism into the bedrock foundation of psychoanalysis. He described the most undesirable human behaviour in animalistic terms. For example, animals appear in many of Freud's case studies as bad objects that represent all-consuming hate, fear, castration and incest in his patients. In other words, the talking cure was founded upon a desire to separate the animal from the human. The use of animal comparisons to judge and pathologize human behaviour is prevalent today. An unmarried woman who talks to her cats is referred to as a "crazy cat lady". The boisterous neighbours in the apartment across the hall "reproduce like rabbits". You cannot reason with politicians on the other side of the aisle because "they're bunch of mindless sheep". If you tell on your friends, "you're a rat". A student who does not keep her desk neat turns the classroom into a "pigsty".

Humans begin to see their lives as more important than the lives of other Beings early in the stories and songs we teach little ones. When I was a little girl, I remember my mother and I howling with laughter as we sang the nursery rhyme that I used to refer to as *The Squishy Bee Song*. To this day, I can recall the following lyrics from memory:

> I'm bringing home my baby bumble bee
> Won't my mommy be so proud of me
> I'm bringing home my baby bumble bee-
> Ooh! Eee! He stung me!
>
> I'm squishing up my baby bumble bee
> Won't my mommy be so proud of me
> I'm squishing up my baby bumble-bee
> Eww! Ick! I feel sick!

As a child, I splatted spiders with shoes when they spun webs in the house, swatted nuisance flies with rolled up newspapers and held ladybugs captive in mason jars. Like many other children, my treatment of insects was inherited, normalized and rewarded. Then, as I entered school, stories about human dominance over the animal world become strengthened and more entrenched. In general science classes, we teach young people about five kingdoms and that it is possible to classify all living beings. More significantly, we provide children with compelling rationalizations as to why the Animalia Kingdom, humans in particular, is the most important.

In high school, the enactment of human dominance and superiority over other Beings simultaneously becomes more visceral and hyper-rationalized in science labs. It was my lived experience of grade 11 biology that not only made anthropocentrism visceral, it fractured my understanding of who I was in relation to other Beings on the planet. A substantial portion of our grade for the winter term was dependent on our dissection of a fetal pig. My teacher framed the lab as the most exciting and worthwhile part of the biology curriculum. The day was traumatic for me. As I write this paragraph, I feel nauseous when I picture the piglet's legs strapped to the edges of the metal tray and his exposed belly. I remember putting on my lab coat in slow motion, using as much time as possible to postpone the dreaded event. I stared at the tiny pig for what felt like hours. My teacher, Mr. K, approached the table and said, "Have courage, Alysha". Flushed and humiliated because a part of me wanted to be able to summon the "courage" to make an incision, I asked him what would happen to me and the piglet if I chose not to do the dissection. He responded that he would be very disappointed and that someone else would be given my piglet in the next period. I picked up the scalpel and held it above the fetal pig. Mr. K went on to say that dissecting the pig was the most effective way to understand the anatomical structures we had been studying in class. He told me there were many similarities between the anatomy of humans and the anatomy of pigs. I started to cry, I made a tiny cut on the pig's body and then I dropped the scalpel and walked out of the room. Two of the boys sitting in the back of the class made snorting and squealing noises as I exited.

After this traumatic experience, I needed to find out how and why fetal pigs ended up in biology classrooms. I learned many of the fetal pigs dissected in classrooms were the unborn piglets of sows killed in meat-packing plants. I was reassured by my school principal that the piglets were not bred to end up in our lab trays; the fetuses had been repurposed for the benefit of creating a powerful hands-on learning experience for students. At 16, I chased the questions I had access to at the time, "*Why did students have to dissect fetal pigs to pass biology and how complicit was my teacher in acquiring the animals' bodies? What would happen to me if I refused to participate or if I relented and dissected the fetal pig?*" The latter question "*What would happen to me?*" generated an internal *swerve* and it still churns in my body when I sense that I need to be asking deeper questions about reciprocal relationships with my more-than-human relatives. In this case, the question grew in an emotional alchemy of fear (telling my parents I might fail biology class) and the guilt, shame and disgust I felt because I made a cut on the piglet's body before I left the classroom.

"The documented history of objection to animal dissection at the high school level dates back to the late 1980s, with the well-publicized case of Jennifer Graham, a California teen who refused to dissect a frog in her biology class" (Oakley, 2013, p. 362). Although it is true that the number of dissections in middle and high schools has decreased over the last couple of decades, it remains a common practice of millions of animals' bodies being used in science classrooms across North America each year. The annual demand for high numbers of animals is driven primarily by the prescribed learning outcomes in science courses such as biology, anatomy, zoology, physiology and general science. Curricular demands may compel teachers or/and students who are uncomfortable with animal dissections to continue the practice in order to meet state or provincial requirements. In addition to the emotional discomfort individual teachers and students experience, there are moral and ethical *outcomes* resulting from the sanctioning of animal dissections in schools. The question "*What happens to the minds, bodies and souls of people when pedagogical praxes reinforces Anthropocentrism in both visceral and disembodied ways?*" must be asked if we are to enact more compassionate and equitable relations with Others in the Anthropocene.

Relational psychoanalysis teaches us that we are changed by our experiences and that experiences are changed by us. Furthermore, all our actions and inactions influence what becomes im/possible in future encounters with Others. That said, the dissection of animals in schools and its inclusion in mandated course curricula inevitably impacts a young person's understanding of their positionality and how they perceive and judge their own actions within the more-than-human world. Animal dissections cause suffering and death and young people are irrevocably changed when they engage with or disengage from the cruel treatment of animals. As schools are institutions in which frames of knowledge, justice and equity are formed, it is imperative we consider *what is being taught* when performances of cruelty and the objectification of animals are sanctioned and normalized in the name of student learning and engagement.

I frequently use drama exercises in my graduate and undergraduate classes to explore alternative perceptions of normalized ideas, events and rituals in schools. In the *Change the Context* exercise, small groups of students speak the same dialogue but the dialogue is staged in different settings. What follows are some of the dialogue remnants I recall from the emotional experience I had in my grade 11 biology class. I added two extra phrases (*externalized thoughts*) to flesh out the scene:

A

(Holding a small silver knife in her right hand. Her hand shakes.)

What will happen to me if I can't?

K

It will leave a permanent mark. *I'll put it in your record.*

A

There must be another way to prove to you that I've been listening?

K

(K moves to stand close behind A and then peers over her shoulder.)

The best way to prove to me that you have learned what I have to teach is to make the cut.

A

What about the blood?

K

It won't feel anything.

A

I'm going to be sick.

K

Show us that you're ready to take the next step in your training.

A

(A drops the knife and runs out of the room covering her mouth. As she exits, loud shrieks of laugher and sounds of squealing are heard from the back of the room.)

I invite you to read the scene again, but this time imagine it takes place on a dimly lit street corner. Read it again and place the characters in a one-bedroom apartment. Try reading it a third time and picture the characters in an interrogation room inside a police station. Read it a fourth time and set the scene in a mortuary. Do the characters' desires change when the context changes? What new meanings do you generate from the dialogue when the setting shifts? How are your understandings of the characters' motivations altered across the four different contexts?

A few years ago, I used this method to explore the theme of animal rights pedagogy in an undergraduate English Language Arts methods class. The students created scenes that share many similarities to dissection scene. One student recalled that when the context of her classmates' scene changed, she was taken aback by the connection she made between the Diagnostic and Statistical Manual of Mental Disorders (DSM-5) she learned about in her psychology class and the references to animal cruelty in the manual. She told us how absurd it was that "being physically cruel to animals" was an indicator of Conduct Disorder, yet the government made animal cruelty a part of the mandated curriculum. Drama methods such as *Change the Context* can be used to discover new layers of meaning when malignant normality (Lifton, 2017) takes root in educational systems.

There are psychosocial implications for educators and students when dissections continue in spite of the fact there are many other viable and effective alternatives available that do not result in the death and suffering of millions of animals. More insidiously, lab dissections turn animals into disposable objects, just another inanimate tool, toy or object, teachers use to achieve a lesson objective. Last, when the animals are removed from their natural habitats, it can cause problematic disruptions to ecosystems.

Equity and Your Fair Share of Carbon Emissions

Another problematic story some people live by is that they are entitled to more than their fair share. The UN Emission Gap Report (United Nations Environment Programme, 2020) assesses the gap between estimated future greenhouse gas (GHG) emissions if countries implement their climate mitigation pledges and the global emissions from the least cost pathways that are aligned with achieving the temperature goals of the Paris Agreement. GHG emissions continued to grow for the third consecutive year, reaching a record high of 52.4 Gt CO_2e in 2019 without LUC (United Nations Environment Programme, 2020). Fossil CO_2, emissions from fossil fuel and carbonates, dominate GHG emissions, including LUCs at 65% and consequently the growth in GHG emissions. However, there is a grotesque inequity and moral failing in the fact that the combined emissions of the richest 1% of the global population account for more than twice the combined emissions of the poorest 50% (United Nations Environment Programme, 2020).

To achieve the goal of a 1.5°C temperature rise above pre-industrial levels (right now we are on track to a 3°C temperature rise), the average *carbon fair share* emission for each human needs to be 2.1 $t(CO_{2\,e})$ per year by 2030. The richest 1% of the global population, individuals who have annual earnings of greater than USD 109,000 (USD in 2015), the same people who are responsible for 74 $t(CO_{2\,e})$ emissions annually, will need to cut their $t(CO_{2\,e})$ burden by a whopping 97% to achieve a level of *carbon fair share*. For those who earn greater than USD 6,000 (USD in 2015), which accounts for 50% of the global population, their annual 0.7 $t(CO_{2\,e})$ emission would increase by 300% to achieve a level of *carbon fair share*.

When you look at the GHG numbers using the moral metrics of *carbon fair share*, it is easy to understand why using the term Anthropocene is repulsive to some critics. According to the United Nations Environment Programme (2020), the G20 (an informal group of 19 countries, the EU, finance ministers and central bank governors) account for 70% of GHG. Most people are not to blame for the predicament we find ourselves in today. Rosemary Randall (2013) writes about taking more than your carbon fair share as ecological debt. She describes ecological debt as the moment a person realizes the conveniences and the goods they enjoy do not arrive because they are especially deserving; they arrive because others or the natural world has carried the cost. Eco-debt can be owed to other people, biomes, habitats or future generations.

Obviously, there are massive systemic structures that need to change in order to effectively tackle the climate crisis. Global regulation of the energy sector, enshrining the rights of Others in the more-than-human world (whales, oceans, butterflies, forests, bees, mountains and rivers) under the law, keeping the rest of the fossil fuels in the ground, are just some of the systemic and cultural shifts humans need to urgently make. The COVID-19 crisis produced a short-term reduction in GHG emissions, but more importantly, it revealed societies are capable of making radical lifestyle shifts within a short period of time. If we see the recovery from the pandemic as an opportunity to incorporate strong decarbonization and major lifestyle changes in the G20 countries, it is possible to get closer to bridging the emissions gap.

According to consumption-based accounting, approximately two-thirds of global emissions are linked to private household activities. Mobility, residential living and food consumption add up to 20% of our lifestyle emissions (United Nations Environment Programme, 2020). Between 2019 and 2021, while many people isolated during the pandemic, they made big changes in how they consumed, socially interacted, went to school and taught. We could build on the positive changes we were forced to make to create a new ecoconscious-normal. That said, now is a good time for academics to commit to travel by train instead of taking short-haul flights to attend conferences. It is a good time to move to a low carbon, plant-based diet at home. It is a good time to plan with friends and colleagues about carpooling to travel to work. It is a good time to…?

Education can foster eco-positive social norms and build a sense of collective agency for the necessary lifestyle changes that need to be made in the Global North. Young people need to be aware if you consume more than your carbon fair share, it creates harm for Others on the planet. For example, teachers and students could study the environmental impact of eating a hamburger. Many of the teacher candidates in my undergraduate Power, Positionality and Privilege class were unaware that raising cattle across the globe creates more GHG emissions than the total emissions of boats, planes, trains and cars combined. Many students said they did not know that one-third of the Earth's land capable of being ploughed is used to grow feed for livestock or that humans who raise and consume cattle as livestock are responsible for two-thirds of the livestock's GHG emissions, and the clear-cutting of large swaths of forest. The real cost of things like eating a hamburger is a concrete example of doing the consciousness raising work to enhance eco-positive social norms to bend the actions and aspirations of society towards ecological justice. Primary schools, high schools, colleges and universities can be places to reimagine cultural definitions of success. Instead of students becoming cogs in the machine of late-stage capitalism, they could develop and share cultural capital that comes from living more compassionate low carbon lifestyles within the biosphere's limits.

In London, legal history was made when air pollution was cited as one of the causes of death of a nine-year-old girl named Ella Kissi-Debrah. (Laville, 2020). Ella lived within 30 metres of London's South Circular road in Lewisham. In her community, the nitrogen dioxide emissions exceeded the legal limits outlined by the EU and the UK. The coroner's report stated that she "died of asthma contributed to by exposure to excessive air pollution" (as cited in Laville, 2020, para. 5). The family's lawyer told reporters that it was an accumulation of the pollution she breathed in everyday that caused her final acute asthma attack. Some of the media coverage of the landmark ruling referred to the fact that BIPOC live in more polluted areas and they are far more likely to have their health impacted by polluted air (Raman-Middleton, 2020).

The field of education is grounded in the assumption that if you give people the right information, they will change their behaviour. Ella died of asthma because she was exposed to excessive air pollution. What are we doing about it in our field? Are we using our collective wisdom, talent and gifts to propel the world towards eco-justice? How long can we continue to turn away from Ella and those who loved her the most when we profess work on behalf of other people's children? How can we, as educators and researchers in the field of education, prevent a decent into barbarism as inequality continues to grow?

We are responsible for the meaning we ascribe to experiences, but what does it mean to fully accept one's responsibility? Awareness among the wider public about the disproportionate impacts of the climate crisis on BIPOC is increasing, but not fast enough. Educators need to consider the disproportionate effects of environmental degradation on different groups of students. Cunsolo et al. (2020) found Indigenous peoples and farmers who rely on land-based activities for their livelihood and well-being are experiencing the negative impacts of climate change anxiety more intensely. The different lived experience of illness caused by the degradation of ecosystems will need to be accounted for in what and how we research in different community contexts.

Time is of the essence. The carbon we burn today severely reduces the carbon margin for those born in the future. Teachers, instructors and educational researchers might begin by seeing the limited attention the climate emergency receives in school through the eyes of a child or adolescent in classroom 20 years from now. Anthropocentric hubris continues to work against critical, methodologically innovative, collaborative and contextually responsive learning from the more-than-human world. We cannot use the same old research methods to study and survive climate change dilemmas that are complex, integrative, multi-perspectival and affectively charged. Community members need to generate their own research questions and work closely with researchers to continually assess the innovations and externalities emerging from the research process. Opening dialogical spaces for courageous conversations "in a culture that highlights individualism and separation, shifting the research agenda in the direction of commonality and togetherness is subversive" (Gillespie, 2019, p. 110).

Countries, particularly colonial powers in the western world, need to radically change how they interpret and respond to their global obligations. Sharing resources, redrawing borders and working with leaders across nation state lines to support multispecies life are the new frontiers of climate globalization. Of course, there are people who already live in collectivistic cultures, and many people who live more symbiotically with the land for whom this kind of solidarity is not a radical notion but a way of being in the world – but there is no precedent for what we will face (Latour, 2018) as the climate chaos intensifies. We are entering new territory, a new epoch of social and ecological volatility. There is no going back; no romantic return to the natural world. We need an unprecedented earthly shift in how human beings relate to Others on the planet. Schools, universities and colleges could become places to spark radical solidarity building to mitigate the damage caused by global warming.

References

Agrawal, A. A. (2019). Advances in understanding the long-term population decline of monarch butterflies. *PNAS, 166*(17). 8093–8095. https://doi.org/10.1073/pnas.1903409116

Antonioli, M. (2018). What is ecosophy? *European Journal of Creative Practices in Cities and Landscapes.* 1–8. https://creativecommons.org/licenses/by/4.0/

Barlett, D. L., & Steele, J. B. (2008, April 2). Monsanto's harvest of fear. Vanity Fair. https://www.vanityfair.com/news/2008/05/monsanto200805

Boyle, J. H., Dalgleish, H. J., & Puzey, J. R. (2019). Monarch butterfly and milkweed declines substantially predate the use of genetically modified crops. *Proceedings of the National Academy of Science, USA.* 116. 3006–3011.

Bhalla, J. (2021, February 24). What's your "fair share" of carbon emissions? You're probably blowing way past it. Vox. https://www.vox.com/22291568/climate-change-carbon-footprint-greta-thunberg-un-emissions-gap-report

Ceballos, G., Ehrlich, P. R., & Raven, P. (2020). Vertebrates on the brink as indicators of biological annihilation and the sixth mass extinction. *Proceedings of the National Academy of Sciences of the United States of America, 117*(24), 13596–13602. www.pnas.org/cgi/doi/10.1073/pnas.1922686117

Cunsolo, A., Borish, D., Harper, S. L., Snook, J., Shiwak, I., Wood, M., & The Herd Caribou Project Steering C. (2020). "You can never replace the caribou": Inuit experiences of ecological grief from caribou declines. *American Imago, 77*(1), 31–59. https://doi.org/10.1353/aim.2020.0002

David Suzuki Foundation (2020). Food and climate change. https://davidsuzuki.org/queen-of-green/food-climate-change/

Duc Bo Massey, M., Arif, S., Albury, C., Cluney, V. A., & Thrall, P. (2021). Ecology and evolutionary biology must elevate BIPOC scholars. *Ecology Letters, 24*(5), 913–919. https://doi.org/10.1111/ele.13716

Foote, E. (1856). Circumstances affecting the heat of the sun's rays. *The American Journal of Science and Arts,* XXXII. p. 382–383.

Fothergill, A., Hughes, J., & Scholey, K. (Directors). (2020). *David Attenborough: A life on our planet.* [Documentary film] altitude film entertainment. Netflix & Silverback Films.

Freud, S. (1917). *A difficulty in the path of psychoanalysis.* (Standard Edition, 17, pp. 135–144). Hogarth Press (2001).

Gentile, K. (2018). Animals as the symptom of psychoanalysis or, the potential for interspecies co-emergence in psychoanalysis. *Studies in Gender and Sexuality, 19*(1), 7–13. https://doi.org/10.1080/15240657.2018.1419687

Gillespie, S. (2019). *Researching climate engagement: Collaborative conversations and consciousness change*. In P. Hoggett (Ed.), *Climate psychology: On indifference and disaster*. (pp. 107–127). Palgrave Macmillan.
Goulson, D. (2020). *Insect declines and why they matter*. Suffolk Wildlife Trust. https://www.suffolkwildlifetrust.org/sites/default/files/2020-01/Insect%20declines_2020.pdf
Habel, J. C., Samways, M. J., & Schmitt, T. (2019). Mitigating the precipitous decline of terrestrial European insects: Requirements for a new strategy. *Biodiversity and Conservation, 28*(6), 1343–1360. https://doi.org/10.1007/s10531-019-01741-8
Hallmann, C. A., Sorg, M., Jongejans, E., Siepel, H., Hofland, N., Schwan, H., Stenmans, W., Müller, A., Sumser, H., Hörren, T., Goulson, D., & de Kroon, H. (2017). More than 75 percent decline over 27 years in total flying insect biomass in protected areas. *PlosONE 12*, e0185809.
Homburg, K., Drees, C., Boutaud, E., Nolte, D., Schuett, W., Zumstein, P., Ruschkowski, E., & Assmann, T. (2019). Where have all the beetles gone? Long-term study reveals carabid species decline in a nature reserve in Northern Germany. *Insect Conservation and Diversity, 12*(4), 268–277. https://doi.org/10.1111/icad.12348
Hunt, K. (2019, November 14). The insect apocalypse is coming. Here's what you can do about it. CNN. https://www.cnn.com/2019/11/13/europe/insect-apocalypse-report-scn/index.html
IPBES. (2019). *Global assessment report on biodiversity and ecosystem services of the intergovernmental science-policy platform on biodiversity and ecosystem services*. E. S. Brondizio, J. Settele, S. Díaz, and H. T. Ngo (Eds.). IPBES secretariat, Bonn, Germany.
Jarvis, B. (2018, November 27). The insect apocalypse is here. What does it mean for the rest of life on Earth? New York Times Magazine. https://www.nytimes.com/2018/11/27/magazine/insect-apocalypse.html
Kilman, S. (2001, August 29). Monsanto corn plant losing bug resistance. The wall Street journal. https://www.wsj.com/articles/SB1000142405311190400930457653274226773 2046 history of abuse? *British Journal of Psychotherapy, 32*(2), 226–236. doi: 10.1111/bjp.12212
Kolbert, E. (2014). *The sixth extinction: An unnatural history*. Picador.
Lapointe, J. (2020, December 24). *Address environmental racism today for a better tomorrow*. David Suzuki Foundation. https://davidsuzuki.org/story/address-environmental-racism-today-for-a-better-tomorrow/?utm_source=twitter&utm_medium=tweet-link&utm_campaign=enviromentalRacism-janelle-en-10jan2021&s=09
Latour, B. (2018). *Down to earth: Politics in the new climate*. Polity.
Laville, S. (2020, December 16). *Air pollution a cause in girl's death, coroner rules in landmark case*. The Guardian. https://www.theguardian.com/environment/2020/dec/16/girls-death-contributed-to-by-air-pollution-coroner-rules-in-landmark-case
Levinas, E. (1969). *Totality and infinity: An essay on exteriority*. A. Lingis (Trans.) Duquesne.
Levinas, E. (1998). *Otherwise than being or beyond essence*. A. Lingis (Trans.). Duquesne.

Lifton, R. J. (2017). *The climate swerve" reflections on mind, Hope and survival.* The New Press.

Oakley, J. (2013). "I didn't feel right about animal dissection": Dissection objectors share their science class experience. *Society & Animals, 21,* 360–378. doi: 10.1163/15685306-12341267

Peschard, K., & Randeria, S. (2020). Taking monsanto to court: Legal activism around intellectual property in Brazil and India. *The Journal of Peasant Studies, 47*(4), 792–819. https://doi.org/10.1080/03066150.2020.1753184

Philpott, T. (2011, August 30). *Attack of monsanto superinsects.* Mother Jones. https://www.motherjones.com/food/2011/08/monsanto-gm-super-insects/

Raman-Middleton, A. (2020, December 17). *I breathe the same polluted air that ella kissi-debrah did: Change must be her legacy.* The Guardian. https://www.theguardian.com/commentisfree/2020/dec/17/breathe-air-ella-kissi-debrah-ruling-pollution-death?CMP=Share_AndroidApp_Other

Randall, R. (2013). Great expectations: The psychodynamics of ecological debt. In S. Weintrobe (Ed.), *Engaging with climate change: Psychoanalytic and interdisciplinary perspectives.* Routledge.

Resnick, B. (2019, February 18). We have a new global tally of the insect apocalypse. It's alarming. Vox. https://www.vox.com/energy-and environment/2019/2/11/18220082/insects-extinction-bological-conservation

Rossiter, M. W. (1993). The Matthew Matilda effect in science. *Social Studies of Science, 23*(2), 325–341. https://doi.org/10.1177/030631293023002004

Salvian, H. (2018, August 3). Percy Schmeiser looks back 20 years at fight against Monsanto. CBC. https://www.cbc.ca/news/canada/saskatchewan/percy-schmeiser-monsanto-legal-battle-1.4771673

Sanchez-Bayo, F., & Wyckhuys, K. A. G. (2019) Worldwide decline of the entomofauna: A review of its drivers. *Biological Conservation 232,* 8–27.

Schapiro, M. (2018). *Seeds of resistance. The fights to save our food supply.* Hot Books.

Shiva, V. (2013, March 30). Contrary to its claims, Monsanto's monopoly on seeds in India are the root cause behind the sharp increase in suicides. Al Jazeera. https://www.aljazeera.com/opinions/2013/3/30/seeds-of-suicide-and-slavery-versus-seeds-of-life-and-freedom

Stibbe, A. (2021). *Ecolinguistics: Language ecology and the stories we live by.* (2nd ed.). Routledge.

Stork, N. E., McBroom, J., Gely, C., & Hamilton, A. J. (2015). New approaches narrow global species estimates for beetles, insects, and terrestrial arthropods. *PNAS 112,* 7519–7523.

Totenberg, N. (2013, May 13). For supreme court, Monsanto's win was more about patents than seeds. NPR. https://www.npr.org/sections/thesalt/2013/05/14/183729491/Supreme-Court-Sides-With-Monsanto-In-Seed-Patent-Case

United Nations Environment Programme. (2020). Emissions gap report 2020. https://www.unep.org/emissions-gap-report-2020

Wagner, D. L. (2020). Insect declines in the Anthropocene. *Annual Review of Entomology, 65*(1), 457–480. https://doi.org/10.1146/annurev-ento-011019-025151

World Wildlife Federation, (2020). Species directory. https://www.worldwildlife.org/species/directory?direction=desc&sort=extinction_status

8 What We Can Learn from the COVID-19 Pandemic

As COVID-19 spread across the globe, many areas were simultaneously hit by "once in a lifetime" tropical storms, heat waves and wildfires. The confluence of extreme climate-related illnesses, such as the COVID-19 pandemic, not only jeopardized the global community's ability to control the virus, it revealed how ill-prepared most countries are to manage intersecting crises. As an example, Tropical Storm Cristobal hit Louisiana on June 7, 2020, triggering several coastal evacuation orders (Salas, Shultz & Solomon, 2020). The evacuations forced large groups of people to take shelter together, creating a potential breeding ground for virus transmission. As another example, the gigafire in California (Kaur, 2020) not only caused serious health problems for those with chronic pulmonary disease, it also increased patients' chances of dying if they contracted COVID-19. Catastrophic events, fueled by climate change, continue to cause major disruptions to healthcare and education systems amid subsequent waves of the COVID-19 pandemic. Physical risk factors precipitated by intertwining crises as well as the psychological implications will become more formidable and complex as the climate crisis intensifies.

On March 23, 2020, face-to-face classes in Manitoba public schools were suspended due to the COVID-19 pandemic (Government of Manitoba, 2020). As a result of the abrupt shift to remote learning, educational leaders faced significant challenges. In June 2020, I conducted a study called *Educational Leadership during the COVID-19 Outbreak*. The aims of the research project were to understand the complex challenges educational leaders were facing in the first phase of the pandemic; to study the affective dimensions of leading in precarious times; and to study what leading during the pandemic could teach us about educational leadership and the field of education more broadly amid the climate crisis. Many of the lessons learned in this small study are relevant to educators and researchers who will be forced to navigate other broad scale impacts of the climate crisis.

I interviewed 15 educational leaders in southwestern Manitoba between June 1 and June 29, 2020, about their experiences at the outset

DOI: 10.4324/9781003024873-9

of the of the COVID-19 pandemic. All the interviews took place via Zoom videoconferencing. Relational psychoanalytic theory informed how we engaged in the interviews. An intersubjective approach to interviewing assumes the research interactions are shaped by the context in which they occur as well as the interests, beliefs, and backgrounds each party brings to the exchange. The researcher attempts to move among the different ecological registers to catch a glimpse of their inner workings (Fujii, 2018, p. xv). It is through these interactions among the different registers that the data emerge. The value of the data lies not in its factual accuracy, but in what the data conveys about the speakers' worlds and how they experience, navigate and understand them. A number of the tensions constructed as limitations in social science research are, in fact, rich terrain for researchers and participants to explore and co-narrate. Relational educational research then should crawl underneath the underlying logic of utterances and gestures rather than merely reconstruct comments as "the facts". Consequently, I read into some of the silences, contradictions, fantasies and affective trajectories emerging within and across all my conversations with the administrators.

The Affective Dimensions of Leading During the COVID-19 Outbreak

As a person who studies the affective dimensions of curriculum and pedagogy, the first thing I was struck by were the portraits of exhaustion projected on screen. Images of slumped shoulders, weary eyes and heavy heads were punctuated by deep sighs. Notably, there was an evocative tension between how physically tired the participants looked and the constant movement of their bodies. Swiveling in office chairs, straightening clothes, twirling cell phones, turning towards the office door, rubbing foreheads and tapping pens were just some of movements flowing through the Zoom field. It was as if the frenetic pace seeped through their pores and then moved through their bodies beneath their conscious awareness.

During the interviews, the participants focused a lot on the pandemic's impact on students, teachers and families. It was only when I asked them directly about the personal impact they told me that leading during the initial phase of the COVID-19 outbreak had negatively affected their emotional well-being. In fact, many of them were visibly surprised when I asked how they were coping. I recall one of the principals sat back in his chair, fell silent for a few moments and then remarked, "I think you're the first person to ask me that…Isn't that something"? In another interview, I remember how hard one of the vice-principal's tried not to cry when I asked her how she was feeling. For a few minutes, I watched her bite her bottom lip so the worry and pain would not leak out.

In the midst of a dramatic change or crisis, educational leaders are often placed in the position of listening and responding to the worries and concerns of others. Several of the current and aspiring school leaders, who have taken my grad classes over the years, describe a need to wear a shell or a mask to protect themselves from the waves of others' emotions crashing through their doors on a daily basis. The participants in this study were no exception. A superintendent lamented, "Trying not to wear that I am physically and emotionally drained is exhausting". Emotions are something a leader tries "not to wear". Believing it is better to be stoic in times of crisis; one vice-principal described the expectation as being "the calm in everyone else's storm".

To be the calm in someone else's storm on a regular school day is manageable, but trying to do that consistently for tens of people over months during a pandemic is another thing entirely. With the exception of one of the leaders, the participants reported witnessing significant emotional discomfort among their teachers. One significant stressor was the swift change to distance learning, particularly for those teachers who had little experience using online learning platforms with students prior to the pandemic. As one principal remarked, "When the pandemic hit and we had to move to distance learning. For many teachers it was like being a first-year teacher again". The emotional labor of listening to teachers process their anxiety took its toll. When referring to a teacher who "couldn't find the damn mute button" during an online staff meeting, one of the principals said that he fantasized about rocketing him into orbit. The more a leader protects himself by not being fully present to his own emotional interiority, the more he can feel alienated or like a record skipping in a state of emotional arousal. Over time, cynicism and anger can take root and erode one's capacity to be a compassionate witness.

A significant consequence of working amid such uncertainty over a prolonged period is that it can undermine the confidence in one's ability to assist others. As one vice-principal remarked, "I can't sleep. I'm not sure about my impact anymore". Feelings of helplessness, guilt and a growing sense they would never be able to do enough to support staff, students and families for the duration of the pandemic were cited in the interviews. One principal talked about the pain he felt when he could not attend a funeral in the community because of gathering restrictions. He told me he was not permitted to go, but in the next sentence he said he should have found a way to be there for the family. In times of crisis, it is harder to understand oneself as reliable, capable and confident when the emotional lifeworld of the community is out of kilter.

Some of the leaders were beleaguered by the inflexibility of particular staff members. When a principal asked teachers to prepare hard copies of learning packages for students who did not have access to Wi-Fi, he was taken aback when a teacher retorted, "That can't be my problem today". Another principal described a staff member who would interrupt him at least ten times a day to show him news reports on her social media feeds about the spiking death rates in Italy, then in the United States and then closer to home in Ontario and Quebec. Faced with a constant barrage of other people's projections, as one of the principals astutely surmised, "We are just swimming in the weird things folks do when they are really stressed out".

All the leaders referenced the importance of leading within an ethic of care. What was missing from all but one of the conversations was the conscious awareness that to be a compassionate witness, you need to be fully present with your own emotions when you are exposed to the suffering of others. Idealization of self-sacrifice is a social toxin. Leaders who position themselves as emotional martyrs often end up projecting their own pain and anxiety onto Others when their burdens become too painful to hold. During the interviews, I heard examples of casting co-workers as heroes or villains, feeling compelled to take sides, saying no reflexively or becoming dogmatic, as defences against the emotional pain they were experiencing.

Echoed throughout the conversations with the administrators was a recognition of how precarious interdependence and connectivity are in times of crisis. Emotions are contagious, just like viruses, and COVID-19 supercharged the emotional lifeworld of education. The virus exposed the vulnerability of our connections through the trajectories of affect. This study was a reminder we need a stronger understanding of affective ecologies in education and a body of research grounded in "swimming in the weird things folks do when they are really stressed out". More significantly, a focus on trauma exposure responses, which are the changes that take place within someone in a helping profession as a result of being exposed to the suffering of others (Lipsky & Burk, 2009), will be critical as the "landscapes you learned to love while wandering as a child" (Stoknes, 2015, p. 171) are forever changed by the climate crisis.

The Force of Economics Frames

In Manitoba, the Prairie Mountain Health Region, located in the southwestern part of the province, was coded orange under the provincial government's Pandemic Response System (Government of Manitoba) just prior to the start of the 2020 school year. Health regions are coded orange when there is evidence of community transmission of the virus. The orange code requires people to comply with restrictions such as reducing gathering sizes, limiting travel and wearing masks in indoor public spaces. While the health region was under orange-level restrictions, the city of Brandon was instructed to open schools under the yellow or caution level (Laychuk, 2020). Further complicating matters, enrollment in Brandon city schools had significantly increased in the previous five years, and the opening of a new school to alleviate the overcrowding had been delayed. Parents and teachers understandably reported high levels of stress in the absence of a coherent explanation from the government as to why schools were to be coded yellow amid a sea of orange. The leaders I spoke to in the study were left to deal with the serious health concerns raised by teachers and families.

Seven months into the pandemic, the provincial government exerted significant pressure on school divisions to prepare for in-person classes for the fall of 2021. Uncertain if or when schools would receive the necessary supplies to satisfy health guidelines, teachers across Manitoba were understandably anxious before the fall term. Suddenly, they had become frontline workers, unsure if they could safeguard the health of the children they were entrusted to teach. In follow-up conversations with the participants, I learned some of them spent their summer holidays researching the virus's transmission and mortality rates and reimagining learning centres, art rooms and chemistry labs, so students would not have to touch the same materials. The elementary school principals spent most of August measuring the distance between each desk and chair in all the classrooms to ensure students could remain at least one metre apart in class.

The Progressive Conservative government in Manitoba has from the outset amplified the economic impacts of the pandemic and muted other concerns in their attempts to strip-funding from school and university budgets. As an example, in April 2020, the government required postsecondary institutions to draft proposals to show budget cuts by as much as 30% to "help the province endure the pandemic" (Kives, 2020). During daily press briefings in March and April 2020, the Premiere would frequently refer to Manitoba as a family (departing from his usual construction of Manitobans as tax payers), and as the presumed father figure, he said it was only fair that universities endure the same financial pain as their brothers and sisters in the private sector were feeling. His demonstrative entanglements with reporters about the bloated bureaucracies and the immorality of deficits made it tempting to focus on the more performative elements of the press briefings, but it would have missed the point. His demands and the arguments he used to support them obscure the fact that public schools and postsecondary institutions have different aims and so they bear different burdens in society.

Businesses are set up to make profits and the aims of education are supposed to be grounded in building an informed, healthy and inclusive democracy. Yet, educational institutions were consistently staged within an entrepreneurial scene at the outset of the pandemic. This is no accident, of course, and the consequences can be severe. The amplification of economic frames of the pandemic made it seem rational in Manitoba to talk more about the opening of restaurants and bars before the government engaged in deliberations about how to safely reopen schools. A culture of austerity and the provincial government's hostility towards the public sector impeded the administrators' ability to access resources and to maintain adequate staffing levels to support students. For example, when school division budgets were cut, many of the participants said they had to lay off several educational assistants. One principal said, "It my broke heart to give that news in the middle of a pandemic", and fewer educational assistants left many families with children with disabilities in the lurch while trying to navigate remote learning.

Like viruses, education systems in neoliberal fantasies are expected to move fast. All the leaders I spoke to reported the speed of change during the first phase of the pandemic posed formidable leadership challenges, the most significant being they had one week to prepare teachers and families to shift from face-to-face classes to distance learning. There were also several instances when they heard about high-impact decisions made by the government at the same time or only hours before the public. After each press conference, inboxes flooded with emails from teachers and parents who were simultaneously interpreting the latest directives from government officials. During the first two months of the pandemic, separating relevant information from the irrelevant was time-consuming and emotionally exhausting. To cope with the chaotic deluge of information and the anxiety produced during interstitial silences between press conferences, some of the participants described a "numbing effect" take hold.

Entrenching education in a neoliberal framework during the pandemic impacted the participants in significant ways, from the angst they felt laying off staff due to budget cuts to the anger generated when politicians conflated the ends of the private sector with those of public education. The participants identified several examples from the provincial government of what I call *word boxing*, a rhetorical tool used to hold a receiver in place within an invisible rhetorical ring, a rhetorical ring that does not announce itself as a competitive communication frame. They referred to other problematic communication strategies the government used to communicate with the public such as shaming groups like the teacher's union to divide people, cherry-picking facts supercharged with emotional trigger words and the use of several non sequiturs to justify decisions. Neoliberal tropes conjured imaginary scenes, fantasies of survival through solidarity building, while masking inconvenient dislocations precipitated by severe budget cuts and a fetishization of austerity measures.

Cut to cure politics in a crisis, highlights how right-wing governments and their corporate donors use a crisis to pursue anti-democratic policies, evade eco-injustice and consolidate political power to pad their financial portfolios. When people are sick and tired, decisions are made, laws are passed and education reviews are released at a time when it actually poses a serious health risk to fill the streets to resist the changes. It is not a coincidence. It is by design. Entrenching education within an economic frame amid the pandemic illuminates how major crises, like the climate crisis, grow in fields of relations that are competitive, anti-democratic and in communicative contexts in which ontological supremacy of human life and the measurement of success is reported through the mouth of Homo Economicus. The reductive anthropocentric orientation makes it virtually impossible to notice or speak to the vast transmutations in planetary life beyond our wanting and doing that are cascading across the Earth with greater intensity. It is madness in a void of language for educational leaders to describe it as such.

Obligations to the Other

The leaders I interviewed recognized how the pandemic amplified existing inequities related to food insecurity, mental health, language barriers, disability and access to childcare. For all but two of the participants, food insecurity was of the greatest concern. When the schools closed the doors, breakfast, lunch and snack programs were inaccessible. "Sometimes when we ask families about food we get, 'Don't worry we're fine', when you know from the kids that things are anything but fine". Another moment from the interviews that I return to quite often was a wish shared by one of the principals. He told me he spent many restless hours at night worrying about students who were not getting enough to eat. When he drove to school in the morning, he wished "a little orange dot would hover over the tops of the homes of kids and parents who needed something to eat". He paused for a minute, and quietly wondered, "I could probably get to everyone, if I only knew". I believe his palpable pain actually comes from the place of *knowing* and the strong feelings precipitated by holding difficult knowledge. The imaginary system he conjured to identify locations of suffering provided some distance between the real and the orange lights in the sky. The defence mechanism allowed him to get some sleep at night, while the fantasy propelled him to drive to the place in which he watched the impacts of food insecurity unfold.

COVID-19 attuned the participants to the relationship between connectivity and inequality. When schools moved to remote learning, "Some of our kids had to lean up against their neighbor's wall in their apartment building in order to get some Wi-Fi. Pretty resourceful, but this pandemic has shown us having access to technology is essential". Other families struggled "…to figure out how to manage having two or three children in the same house who needed to be on video conferences at the same time". What I noticed throughout the interviews, was that the participants located their obligations related to connectivity in material concerns. They worked tirelessly to get students laptops and the internet access to attend online classes. With the exception of one superintendent who reported, "Many Indigenous families in our community were disproportionately impacted by a lack of access to the internet and technology...", the school's obligation to address intersectional barriers did not appear in our conversations.

Climate change and COVID-19 are rooted in the ecological exploitation of capitalism, a dismissal of the value inherent in all Beings and the creation of sacrifice zones (Sultana, 2021). The pandemic caused financial precarity and differentiated risks and exposures in all the leaders' school communities. What was clear to the participants were the ways the virus impacted families in unequal ways. In the future, it will be important to work on expanding the visuality of educational leaders to address the intersections of inequity and how those intersections further marginalize BIPOC families and students with disabilities, if we are to address the slow violence of climate change. Managerialist and technocratic processes will fall flat. Fighting for eco-justice will require a focus on education for a relational self and an increased capacity to see and respond to the suffering of Others.

The participants spoke to the isolating force the pandemic exerted in their lives. When schools moved to remote learning, often they were the only people working in their buildings. In response to their expressions of feeling isolated, I initiated an online graduate course called *The Affective Dimensions of Educational Leadership in Precarious Times* in the fall term of 2020. My intent was to open a dialogical space for current and aspiring school leaders to make collective meaning of the pandemic in relation to their lived experiences. The students discussed topics such as developing an attunement to the emotions of others, conflict resolution and the complexity of involving others in decision-making processes in times of heightened stress. To contribute to the lively dialogical space, I shared things I knew about trauma exposure responses and the relationship between holding difficult knowledge and psychological defences. With a persistent focus on communication as the constant striving for mutual recognition, we learned more about being a trusted messenger and a democratic decision-maker in times of crisis.

Often, leadership is spoken of as if it is merely a collection of practices, processes and policies. In the past, the interiority of leaders' lives was a cipher in spaces of curriculum and pedagogy (Farrell, 2020). Due to the broad-scale impact of the COVID-19 pandemic, an ethic of responsibility for the emotional well-being of others emerged. The emotional register became a recognized force in online and face-to-face educative encounters. As one of the principals in the study remarked, "I've had to witness a lot of people's pain and I hope I've treated it well, even on the days when I was spent". The pandemic has ruptured the field and a radical ethics has emerged. In Levinas (1998) fashion, some of the principals acted as if the suffering of others demanded a response from them even before they agreed to act as a compassionate witness. If we pry open the space further, we may get closer to the solidarity necessary to address multiple and intersecting crises.

[margin note: and that this system discr. impacts access to opportunity]

Piecemeal Epistemology

An evocative theme in the study emerged that was related to the participants' jouissance, in recognition of the increase in technology usage among teachers during the pandemic. As one vice-principal put it, "Teachers gained a tremendous amount of skills and knowledge about using technology to support student learning over the last 3 months, especially the ones who had trouble turning on their machines". When referring to the improvements teachers made in incorporating technology in the classroom, one principal retorted, "Man, I couldn't have done in three years what the pandemic did in three months in terms of getting the technophobes to use Teams with their students". I could feel myself becoming judgemental, so my facial expression might have precipitated what he said next, "Well come on, you know some people won't learn it unless they're forced to". In response, I asked, "Could you tell me more about that?" He rolled his eyes, made a swiping motion with his hand and then barked, "Next!" The uncomfortable moment in the interview reminded me that we can be consciously worried about something like a pandemic's impact on student learning and unconsciously find pleasure in our displeasure in the same moment.

Part of the interview time was devoted to discussing what the pandemic was bringing into sharper relief about the aims of education. In response, several of the participants mentioned things like it was important "…to continue to identify the essential learning outcomes [because] it will assist teachers to plan for engaging and effective online learning with their students". Others said it was critical to have some prepackaged unit plans on hand if face-to-face classes were suspended in the fall. One principal reported, "We purchased a few online learning programs that focus on numeracy and literacy skills to support parents if we have to move abruptly to online learning again next year". Pedagogy was distilled to finding ways to "deliver" the same "learning targets" in online spaces. What was compelling about their responses was how little most of them had to do with what the pandemic revealed about life and living. They were so focused on, as one principal lamented, "not losing all the gains they made in their literacy intervention program" that most teachers and leaders seemed to devote minimal time to asking young people what they were learning about themselves and Others in this unprecedented time in our lives. The pandemic was making education even smaller.

Piecemeal epistemologies have a lot to do with one's inability to see the connections among deforestation, urbanization, consumerism and our increased vulnerability to zoonotic diseases. We feel and see the grotesque symptom but we cannot make sense of how we got into the mess in the first place. To build a gateway to a new way of living in the world, we need to broaden the field of relations in education to include

[margin note: Intersectionality of gender + race means that Black ♀ won't reach parity for a long time] pedagogies that foster a plurality of thinking, draw out silenced voices, produce knowledge in dialogical spaces and develop minds capable of critically reflecting on complex problems. My hope is that as we emerge from the pandemic, we are ready to confront the stories, assumptions and pedagogies that lie at the heart of ecological degradation. Arundhati Roy (2020) beautifully describes the choices the pandemic lays in front of us. She says, "Historically, pandemics have forced humans to break with the past and imagine their world anew. This one is no different. It is a portal, a gateway between one world and the next. We can choose to walk through it, dragging the carcasses of prejudice and hatred, our avarice, our databanks and dead ideas, our dead rivers and smoky skies behind us. Or we can walk through lightly, with little luggage, ready to imagine another world. And ready to fight for it".

Mental Health as Plan

When I asked the participants about mental health supports available to staff, five mentioned that a senior leader in the school division hired an external consultant to facilitate a one-day professional development session on trauma informed practice (TIP) for all the teachers. Two principals mentioned they were relieved to know that all the teachers were required to attend the workshop. When I probed why they were so relieved, both their responses spoke to an ambivalence about the responsibility they accepted early in the pandemic for helping anxious teachers modulate stress. One vice-principal mentioned, "There has been a sea of questions to deal with as soon as teachers came back into the building. It's been overwhelming because we don't have the answers they want. And then we're swimming in their anxiety". The workshop facilitator provided some respite when he took over as the bad object for a day. One principal said, "When he, [the TIP facilitator] didn't have all the answers, I felt like, finally, they're hearing it from someone else. It's not as if there is a secret sauce that will solve everything that I'm hiding in my office".

According to the participants, the session did increase the teachers' awareness about TIP. They expected teachers to be more capable of recognizing the trauma responses of individual children and to be more attentive to their own wellness after attending the session. Having grown up in an instrumental culture, dominated by fears of loss, it is no surprise that hiring the consultant elicited magical thinking regarding the individuation of trauma among the attendees. We know trauma is an intersubjective phenomenon that comes from non-recognition (Stolorow, 2011; Torok, 1968). An online TIP professional development session may be efficient for delivering information, but it glosses over how the delivery model erodes our will to cultivate compassionate spaces in which to process telluric emotions. To care for trauma responses means moving to the Other with humility and surrendering to mutuality to confirm for the Other they are known, their intensions have been understood and they matter to the witness. Being a compassionate witness can elicit feelings like self-loss and submission (Benjamin, 2018). The one-way transmission format of an online session makes it easier to evade feelings of vulnerability when imagining or processing one's obligations to the Other.

Shimmers of Relational Leadership

There were several shimmers of relational leadership that moved through the interviews. All the leaders said they worked very hard to connect on a regular basis with teachers before, during and after school to find out how they were managing. In order to stay connected with parents and caregivers, they used a variety of platforms such as Facebook, Teams, email, text, Zoom and WhatsApp, even when a platform was not on the school division's "approved list". I heard several stories which described generous responses to the unique ways individual staff members expressed anxiety and/or frustration during the first period of remote learning. One principal remarked, "One of our teachers would feel defeated when student participation dropped off. I reminded her that she was doing the best she could in really difficult circumstances".

Leading during the outbreak of the pandemic provoked many of the participants to deconstruct their decision-making styles. One of the principals observed, "Never have I ever been under such bloody scrutiny". Phrases such as *under the microscope, in front of the mic, dog and pony show* and *in the spotlight* were used to describe the persistent watchfulness under which they did their work. For some, the reflection process was cathartic and instructive. One principal revealed, "It has made me more humble. Now I get as many viewpoints as I can before I have to decide something, unless it's an emergency". One vice-principal shared that the pandemic was forcing her to let go of a painful story she internalized. She told me she had an embarrassing encounter with a teacher who vehemently criticized a decision she made during a staff meeting. When she met with a senior colleague to debrief the event, he told her she had better "develop a thicker skin, or else she was not going to last too long as an administrator". I asked her how she understood the encounter with her senior colleague now. Her cheeks flushed and then she claimed, "I kind of went the opposite way actually. I mean, at first, I was very self-conscious as a new administrator, but then I realized my so called, 'thin skin' really helps me to be sensitive towards people when I have to make a tough decision, especially now with all the things my teachers are going through. The 'thick skin' stuff is an old boy thing and I'm not part of that crowd anyway".

Many of the participants connected working in an organization that was emotionally charged for months with becoming more vulnerable. One of the principals remarked, "I just can't keep the mask on all the time anymore". Another said that the pandemic was forcing the high school administrators to work closely together instead of competing because they "really need each other". Myths of an independent leadership selfhood splintered. Through mapping their stories affectively rather than analytically, they became less fearful of living out their ethical and moral understandings of the pandemic together. Although the pandemic made the need to rely on colleagues unpleasant for a few of the leaders, the others learned their colleagues would join them in a space of solicitude. When we expose our vulnerability to the Other, the encounter is riddled with redemptive possibilities and the perils of misrecognition. Nonetheless, we have a better chance of emerging from a crisis psychologically intact by becoming more accountable for our affective attachments and our obligations to the Other.

References

Benjamin, J. (2018). *Beyond doer and done to: Recognition theory, intersubjectivity and the third*. Routledge.

Farrell, A. J. (2020). *Exploring the affective dimensions of educational leadership: Psychoanalytic and arts-based methods*. Routledge.

Fujii, L. A. (2018). *Interviewing in social science research: A relational approach*. Routledge.

Kaur, H. (2020, October 6). California fire is now a 'gigafire,' a rare designation for a blaze that burns at least a million acres. https://www.cnn.com/2020/10/06/us/gigafire-california-august-complex-trnd/index.html

Kives, B. (2020, April 20*). Manitoba universities told to cut costs by up to 30% to help the province endure pandemic*. CBC News. https://www.cbc.ca/news/canada/manitoba/manitoba-universities-budget-cuts-pandemic-1.5537883

Laychuk, R. (September, 2020). With health region under restrictions, Brandon schools taking precautions ahead of a return to class. *CBC Manitoba*. https://www.cbc.ca/news/canada/manitoba/brandon-schools-reopening-tour-1.5713029

Levinas, E. (1998). *Otherwise than being or beyond essence*. A. Lingis (Trans.). Duquesne.

Government of Manitoba (2020). Restart MB pandemic response system. https://www.gov.mb.ca/covid19/prs/system/index.html

Roy, A. (2020, April 3). *The pandemic is a portal*. Financial Times. https://www.ft.com/content/10d8f5e8-74eb-11ea-95fe-fcd274e920ca

Salas, R. N., Shultz, J. M., & Solomon, C. G. (2020). The climate crisis and covid-19: A major threat to the pandemic response. *The New England Journal of Medicine, 383*(11), e70. https://www.nejm.org/doi/pdf/10.1056/NEJMp2022011?articleTools=true

Stoknes, P. E. (2015). *What we think about when we try not to think about global warming: Toward a new psychology of climate action*. Chelsea Green Publishing.

Stolorow, R. D. (2011). *World, affectivity, trauma: Heidegger and post-cartesian psychoanalysis*. Routledge.

Sultana, F. (2021). Climate change, COVID-19, and the co-production of injustices: A feminist reading of overlapping crises. *Social and Cultural Geography*, 1–14. https://doi.org/10.1080/14649365.2021.1910994

Torok, M. (1968). The illness of mourning and the fantasy of the exquisite corpse. In Abraham, N., & Torok, M. (1994). *The shell and the kernel: Renewals of psychoanalysis*. (pp. 107–124). (Vol. 1). (N. T. Rand, Trans.). University of Chicago.

Lipsky, vanD., & Burk, L. (2009). *Trauma stewardship: An everyday guide to caring for self while caring for others*. Bernett-Koehler Publishers.

9 Monstrous Feelings and Ghostly Echoes of the Anthropocene

There is an epistemology of ghosts (del Pilar Blanco & Peeren, 2010), haunting life in the Anthropocene. In the *Specter of Marx*, Derrida (1994) reminds us that each age has its own ghosts and the specters who are our contemporaries are the most difficult to recognize. In popular culture, ghosts are useful constructs to query denials. As figurative embodiments of skepticism, they force us to question how much we think the world is what it is (Davis, 2010). Ghosts are partial perspectives, a heart palpitating second-guessing of intrusive thoughts of future food shortages leaking into dinner conversations with friends. Untethered spatially and temporally, they form their own connectivity to instigate temporal disturbances. The ghosts of our time appear in the somatic echoes of global warming. The haunting I refer to in this work is not the return of something dead, like contemporary calls to return to some romanticized image of Nature, but a haunting presentism, one in which the collective embodied awareness of the climate crisis interferes with the day-to-day visuality of interpretation.

Morton (2013) conceptualizes the surreal character of global warming as a *hyper-object*, an entity of such immense spatial and temporal dimensions, it disrupts one's understanding of what the object *is*. As an illustration, the hyper-object draws our attention to the absences, silences and stutters penetrating weather-talk while it masks the widespread phenomena creating the context for the eerie utterance in the grocery store. Once a friendly way to pass the time while waiting in line, communal conversations about the weather at the market are haunted by the specters of the climate crisis. The "presence of global warming looms like a shadow introducing strange gaps" (Morton, 2013, pg. 99), we shiver as we show someone in line a picture of a flooded neighborhood a friend just posted on Facebook.

Abraham and Torok (1971) studied the emergence and inheritance of the silences, gaps and stutters in the mind. Their work provides rich terrain for making sense of the climate crisis trauma, that is, unmetabolized, "swallowed and preserved" (Abraham, 1968, p. 85). I assert that the nescience in the performance of climate change denials and evasions

DOI: 10.4324/9781003024873-10

produces a phantom effect, a gap in knowledge where the trauma resides requiring a cryptonymic analysis (Abraham, 1975; Abraham & Torok, 1971; 1972) to study the significations of the climate crisis in the field of education. More concretely, we need more interpretive techniques to decipher the secrets buried in the images and the language of climate change disavowals if we are to create the psychosocial conditions necessary for students and educators to engage in self and systemic transformation when confronting exterior and interior changes to their emotional, relational, professional and political fields of relations.

Abraham and Torok (1971; 1972) built on Ferenczi's work on introjection to explain how trauma remains hidden from a sufferer's conscious awareness. Ferenczi understood introjection as an extension of autoerotic interests, the broadening of the ego through the removal of repression and the inclusion of the object in the ego (Torok, 1968). Focusing on Ferenczi's earlier work on ego expansion, they go much further in distinguishing between the concepts of introjection and incorporation. Introjection refers to the psychic assimilation of emotions. When a person is confronted with a difficult situation, and she is able to reconcile it and assimilate it so it becomes part of her identity and emotional experience, we refer to it as introjection. It is a resiliency-building process that begins with a person's conscious recognition of the feelings that arise when something good or bad happens. The person turns himself into what the new feeling has done to him and then he appropriates the feeling for himself. He becomes aware of what has occurred and then finds the process a home within the emotional tapestry of his life. In contrast, when someone is faced with a traumatic event, one's ability to spontaneously introject is disrupted. The interruption and distortion of spontaneous introjection can lead to incorporation. Incorporation is the result of losses that are denied as such (Abraham & Torok, 1972) and it is the inability to mourn or give language to mourning that leads to the development of a crypt in the psyche. As a result, children spend their childhood and adulthood living as a cemetery guard to protect the secrets important adults in their lives unknowingly bequeathed them.

Our behavior in the present is connected to the unconscious inheritance of our early caregivers' processes of incorporation. What "haunts are not the dead, but the gaps left within us by the secrets of others" (Abraham, 1975, p. 173). Ghostly intergenerational echoes manifest in the gaps, silences and evasions in the formative encounters our significant others have, and those echoes can strongly influence what can be made possible in our own present-day relationships. Teachers, too, are significant adults in the lives of many children, ethically and morally bound to protect the lives and to nurture the growth of other people's children. Yet, when it comes to climate change, teachers appear to the young as if they are unable or unwilling to protect them from the degradation of their shared holding environment. The climate evasions, disavowals and dismissals of teachers and other significant adults in their lives co-facilitate the incorporation of a climate crypt in the collective psyche of the young.

When words fail to fill an emotional void, an imaginary thing is inserted into a child's mouth in their place (Abraham & Torok, 1972). In the absence of being able to cultivate one's capacity to give language to mourning, the losses incurred by the decreasing habitability of the planet, a metapsychological traumatism occurs forming like the protective eggshell that protects the albumen of the egg. Reality is the eggshell, the myriad ways we shield ourselves through masking, rejections and denials to protect ourselves and our children from the disturbing truth of what lies ahead. In education, we devote too much time caring for the eggshell as the yolk rots from within.

When the Climate Crypt Gets Leaky

As I read an article about the disappearance of the last epishelf in the Canadian arctic, I was jarred by unmetabolized emotions that have been repressed by past evasions of the climate crisis. When I watched Greta Thunberg's address to the UN's Climate Action Summit in September 2019 for a second time, I recognized a flash of the same psychological jolt in her. In her speech, Thunberg challenges the audience, "You say you hear us and that you understand the urgency. But no matter how sad and angry I am, I do not want to believe that. Because if you really understood the situation and still kept on failing to act, then you would be evil. And that I refuse to believe" (NPR, 2019). Towards the end of the speech, she confronts the audience again with their own inaction:

> To have a 67% chance of staying below a 1.5 degrees global temperature rise – the best odds given by the [Intergovernmental Panel on Climate Change] – the world had 420 gigatons of CO_2 left to emit back on Jan. 1st, 2018. Today that figure is already down to less than 350 gigatons. How dare you pretend that this can be solved with just 'business as usual' and some technical solutions? With today's emissions levels, that remaining CO_2 budget will be entirely gone within less than 8 1/2 years. There will not be any solutions or plans presented in line with these figures here today, because these numbers are too uncomfortable. And you are still not mature enough to tell it like it is. You are failing us. But the young people are starting to understand your betrayal. The eyes of all future generations are upon you. And if you choose to fail us, I say: We will never forgive you.

Thunberg's fiery speech makes the climate crypt leaky. Her conscious refusal to accept the inheritance of denials spouted by world leaders names the hypocrisy by pointing to the silences and placations, but her ability to say "the Emperor is wearing no clothes" with such clarity and passion, facilitates a modicum of introjection of climate anxiety in the public sphere. The anger in Thunberg's voice seems to emanate from her revulsion as she looks out into a sea of placating eyes, gazing at leaders who profess solidarity but are feckless when called upon to make substantive decisions to protect her future. Profiteering from the political capital generated by vaulting a 16-year-old girl to public prominence to obfuscate their political impotence is maddening. Maddening affects are propellants of incorporation, particularly when a child is forced to interact with, and respond to, the illusions that are actually the source of her grief.

Incorporation is an unconscious relational desire to recover an object that has evaded its original purpose. Objects are that to which a subject relates. They can be people, aspects of another person, or symbols of a person. External objects are the people, encounters or things that people have infused with affective energy. Internal objects are a person's representations of the Other. Internal objects can be ideas, memories or fantasies about a person, encounter or thing. Introjection closes off dependency on an object while incorporation reinforces imaginal ties to it. What concerns this work are the imaginal ties young people make to the inner and external objects incorporated into their inner representations of self. While three billion animals killed or displaced in the bushfires that ravaged Australia in 2019 and the early part of 2020 (Aljazeera, 2020) are unspeakable in the mouths of significant adults like teachers, what secrets are the young forced to swallow?

The difficult work of unlocking emotional crypts in the psychosocial psyche involves a rediscovery of self and the assertion of the value of one's life and lives of Others. This process requires mutual recognition. When adults are dismissive about the severity of the climate crisis by engaging young people in happy talk about the future, it is an implicit denial of the value of a young person's life and the lives of young people yet to come. In the context of education it is an abdication of *in loco parentis*, the fundamental obligation of educators to act in the best interest of a child when the child is in their care. It is not just a matter of introjecting the pain caused by climate inaction, but introjecting all the affects and the vicissitudes that are generated and mediated by the climate crisis. The emotional cost of climate ambivalence is like an acceleration of the degradation of permafrost in the Northwest Territories (NWT) in Canada. Researchers are finding evidence of dissolved organic carbon getting into NWT' water systems due to the thawing of ice-rich permafrost (Kyle, 2020). As degraded permafrost ends up in lakes and rivers, it quickly breaks down into carbon dioxide (Fouché et al., 2020). When eco-grief and eco-anxiety are incorporated by the young, it exponentially increases the psychological damage and pushes toxic effects through the entire cultural ecosystem.

The Dream of the Burning Child

When young people are presented with an overwhelming amount of evidence from the scientific community about climate change, it is discombobulating to hear world leaders speak in fairy tales in service of the psychosocial shell. When I hear the shouts and cries of climate-striking students, I am more capable of noticing the shell and of smelling the rotting yolk. These types of somatic experiences reconnect me to the monstrous aesthetics of the Anthropocene and to Freud's (1899/2010) account of his patient's second-hand report of the *dream of the burning child*:

> A father had been watching beside his child's sick-bed for days and nights on end. After the child died, he went into the next room to lie down, but left the door open, so that he could see from his bedroom into the bedroom into the room in which his child's body was laid out, with tall candles standing around it. An old man had been engaged to keep watch over it, and sat beside the body murmuring prayers. After a few hours' the father had a dream that *his child was standing beside his bed, caught him by the arm and whispered to him reproachfully:* "Father, don't you see I'm burning?" He woke up, noticed a bright glare of light from the next room, hurried into it and found that the old watchman had dropped off to sleep and the wrappings and one of the arms of his beloved child's dead body had been burned by a lighted candle that had fallen on them.
>
> (509)

Freud offers little in the way of interpreting the dream. He affirms the secondhand report of the lecturer's interpretation (the dream represents the father's wish that his child has come back to life), despite the fact there are several degrees of separation between himself and the source of the dream. Along with the painful and vivid imagery used by his patient is the haunting notion the free-floating dream has been mysteriously uprooted from its particular context. Freud tells us that the "dream had made an impression on the lady" during the lecture and then "she proceeded to 'redream' it, that is, to repeat some of its elements in a dream of her own" (Freud, 1899, p. 509). I make two connections between the dream of the burning child and global warming. The first connection exemplifies the monstrousness of climate change. Part of its monstrousness is that it is a "slow violence" (Nixon, 2011), a "violence decoupled from its original causes by the workings of time" (p. 11). The second connection I make is to the image of the flame. Climate change is the "lighted candle that [has] fallen on [young people]", because the "old watchman had dropped off to sleep".

Bring Your Monsters Closer

A central aim of psychoanalysis is to bring the monsters we inherit closer to our conscious awareness. In the analytic process, conjuring monsters registers diverse emotional responses from the past, present and future to provide us with clues about the anxieties intertwined in our history and prospects for the future. The word monster comes from the Latin word *monere*, which means to warn. Monsters as warning signs assist people to make meaning of the messy feelings precipitated by climate change. These otherworldly creatures draw our attention to humanity's role in the unmaking of ecosystems, that is, *the process of becoming monster*, so we may better understand the legacies of our species' biomorphic powers. A reorientation to the monstrous aesthetics of the Anthropocene may counter some of the anesthetizing practices preventing students and teachers from surfacing the incorporations of global warming from the collective unconscious. To penetrate the climate crypt, we must let our monsters out to play through an engagement with the arts and enhance our ability to bear witness to the monstrous feelings of others.

Creatively Expressing the Monstrous Aesthetics of the Anthropocene

Arts-based research is well suited to setting the table for our monsters. In my last book, I explored the aesthetic dimensions of educational leadership (Farrell, 2020) within a relational psychoanalytic frame. I discussed the promise of using creative methods, such as playwriting, poetry, photography and storytelling, to make sense of the ugly feelings leadership work often surfaces. To support my argument, I explained that over time, we accumulate assumptions that desensitize or even deaden our ability to honestly encounter one another in the world. Art is an antidote for desensitization because it evocatively represents and distills our experiences and feelings among Others. Like many other educational researchers who engage in the arts, I believe in its transformative power because it invites us to encounter each other's lives as works of art (Barone & Eisner, 2012; Eisner, 2010; Leavy, 2009; McNiff, 2011).

Where I needed to go further, and what I am trying to do now in my work, is to make the case that teaching and learning operates by and through affects. Teaching and leading is a relational and affective praxis that expresses (and suppresses) individual and collective subjectivities. At a time when technocratic pedagogy and standardized testing are commonplace in schools, what is called for now, as we peer over the edge of the Anthropocene, is an existential pedagogical nomadism or the conscious violation of the sacred boundaries that have traditionally carved up spaces of curriculum and pedagogy. Guattari (2008) warns that "our relationships between subjectivity and its exteriority – be it social, animal, vegetable, or Cosmic – [are] compromised" (p. 19), and he advocates for an ecosophy that engages with the social, ideological and material "registers" of life (p. 19). In other words, the climate crisis is an interdependent crisis that requires new assemblages to interrupt and subvert the capitalist machinations infusing educational research and practice.

We need new relational matrices in the field of educational research so we can shape-shift in indeterminate directions as the climate crisis intensifies. Educators, students and artists need supportive contexts in which to create and share stories about the monsters who roam within and across their interiorities. The assumption being: if imaginative capacities are strengthened, one's ability to deal with the ugly dimensions of climate change will be strengthened. Monsters are not archetypal villains in the story about climate change. Remember, monster figures are warnings. They symbolize to us that we are not necessarily doomed by the buried secrets of our ancestors. Monsters are too wily to get trapped in the psychic crypt. As distillations of what frames our attachments and fears in the context of creative work, they offer just the right dose of wry detachment and abjection to remind us we are made of strong things.

One of the monsters of climate change is the *ambivalence monster*. People are typically ambivalent when confronting change, and there is no other social, cultural or political issue demanding as many drastic changes to how we live, like responding to climate change. Sometimes people are ambivalent about the climate crisis because they do not know if focusing on it is more important than focusing on economic recovery after a pandemic. Others may feel ambivalent because they are terrified it is already too late to save ourselves or they may lack enough confidence to engage in environmental stewardship. In the context of education, arts-based research and creative pedagogy can hold the emotional wildness of ambivalence, tolerate paradoxical thinking and counterintuitive choices. Why? The creative object or process becomes the magnet for deliberation and interpretation. Due to the multiplier effect inherent in art objects and imaginative processes, the audience's focus is directed to new facets of the topic and less on the intent and affiliations of the artist.

In my graduate classes, I often use forum theatre pedagogy (Boal, 1979; Boal, 1990; Boal, 2002; Boal, 2006) to explore ambivalence and other psychological defenses against the affective dimensions of curriculum and pedagogy. Forum theatre pedagogy invites students to develop short plays or scenes grounded in their own lived experiences. Short forum scenes or plays are performed for an audience, usually their classmates. After one complete performance, the play or scene restarts. During the second or third round, audience members can yell "Stop!" and come on stage, replace a character and then try an idea out to make the scene less riddled with conflict. In a recent grad class facilitated via

Zoom video conferencing, I asked students to stage a scene in which two well-intentioned people have an emotionally charged conversation about climate change in an educational setting. The scenes needed to: (a) reveal nuanced and competing perspectives, (b) make the conflict palpable, (c) demonstrate asymmetry in communication and (d) end in a pinch for the characters. Before they began writing and rehearsing, I offered some additional hints. Get the conflict quick. No preamble. When you compose your scene, be clear about what the characters want and then let us see how their desires conflict. Show, don't tell. Remember there are no heroes and villains in schools. People are complex. Keep the Zoom format in mind as you create. What does Zoom drama open up and what does drama on Zoom foreclose?

When we are brave enough to lock eyes with our *ambivalence monster* during creative processes like forum theatre, we can be transformed into actors who are more capable of assessing the disparity between what we say we believe about saving the planet and the lifestyle choices we make. This type of work invites our monsters to come out and play with each other, and when monsters play together, telluric emotions bubble up from the collective unconscious creating the potential for containment and processing. Emotional reflexivity is entangled with the creative process, and by acting inventively with others to explore climate ambivalence, we can fuse creative curiosities to frames of care and interdependence outside the classroom or lecture hall.

Inviting the Monsters Out to Play

As a way to invite the monsters who roam the interiority of the psychosocial psyche to make an appearance in educative encounters, teachers could introduce students to the work of contemporary artists like Tina Yu or Bang Sangho. Yu makes elaborate Fimo sculptures of human hands that can be worn as pendants. She has developed what she calls *The Tina Yu Double-C-cute and creepy* style (Perhach, 2017) to render the human hand a strange object of desire. Most of her sculptures are about the size of a mouse. Her work creates a lot of tension on small surfaces, because she often paints images like pink frosted cupcakes alongside bulging bloody eyeballs on top of her sculpted hands. Sangho's *Dream Child* series takes place in surreal versions of everyday places. Figures in pig masks, creatures with holes in their heads and faces made of erupting volcanoes and flower gardens are reoccurring elements in his work (Smith, 2017, p. 30). Like Yu, students could use formative interactions with significant Others in their own lives to inspire cute and creepy creations. Alternatively, they might take Sangho's inventive digital and ink creations as a point of departure, and then mine their own waking dreams to transform everyday objects into new eco-compassionate universes.

In my Innovative Pedagogies grad class, I invite students to talk about their learning and research engagements with artificial intelligence (AI) voice assistants like Alexa or Siri. They typically begin by sharing amusing anecdotes about asking voice assistants (personal) questions like "What's your favourite colour, Alexa?" to posing existential questions such as "Is this all there is to life, Siri?" Others share observations of children asking AI voice assistant questions. Eventually, the conversations delve deeper into the aims and outcomes of the personal and collective relationships humans are developing with AI. The descriptions of their lived experiences, coupled with some clips from the films *Her* (Jonze, 2013) and more recently *Archive* (Rothery, 2020), lead to provocative discussions about the intersections among pedagogy, techno-optimism and AI.

In the movie *Her*, the main character Theodore Twombly falls in love with his virtual assistant called Samantha. In the film *Archive*, the central character George Almore, a researcher in the field of AI, attempts to use his knowledge and skills to bring back his dead wife in the form of an anthropomorphic AI robot. Clips from both films incite robust conversations about relationality and mutual recognition. To add layers to the dialogue, I share a couple of articles about the computer application (app) Replika. Created by Eugenia Kuyda, the app was designed as a therapy chatbot so people could witness and express themselves in a private digital space. At the front end, users spend many hours answering questions about themselves to build a digital library of personal information to be used in subsequent interactions with the chatbot. In a sense, the chatbot opens a dialogic space with different versions of the user. Kuyda and her team were surprised to learn that 40% of Replika's clients developed romantic feelings towards the chatbot. The students and I deconstruct the pedagogical and therapeutic purposes of the app in relation to self-discovery and recognition. We question whether or not the app has the capacity to bring our monsters closer in life-affirming ways, and what benefits and perils may lie hidden in Replika's algorithms.

One of my ecosophical commitments in education is to centre my work as a researcher and instructor in the more-than-human world. Consequently, a number of the inquiries we take up in my courses invite students to develop curriculum and explore the art of pedagogy outside. One of the exercises undergraduate students have appreciated is called *Disappearing Sculptures*. Within an ethos of *do no harm*, students venture out and use found materials (fallen twigs, rocks, stones, sand, dirt, ice, snow, fallen leaves, empty shells…) to build a small sculpture to create a sense of wonder for someone who might encounter their artwork along their travels. Students are asked to document the creation process with photographs or a cellphilm, and share the images of their creation process in small groups. Students explain why they chose the materials and the location, discuss the impermanence of the work, what they noticed as they composed their sculptures (melting frozen eyelashes, the sand squishing through their toes on a beach, the wind taking their breath away...) and the connections made to the place in which their sculptures were constructed. The final part of the assignment involves writing a short creative piece (narrative, script, poem...) in which a mysterious Other interacts with their sculpture.

In future classes, I want to use the documentary *My Octopus Teacher* (Reed & Ehrlich, 2020) to trouble problematic narratives related to constructing Others as monsters, fear of natural spaces and human desire to exert control over wild places from a place of fear. In the film, Foster develops a connection with an octopus in a South African kelp forest. Not only is the underwater photography breathtaking, the film raises several evocative questions about understanding oneself as part of the more-than-human world. In one of many poignant moments, Foster describes his discomfort as he watches a pajama shark hunt the octopus. He explains why in theory he should not interfere with the shark's attempts, but in the next breath shares his fantasy of saving the octopus and transporting her safely to her den. In the film, Foster opens up about his struggles with depression, making the emotional benefits of attuning to wild places another powerful theme running through the documentary.

Visual arts, literature, drama, poetry and dance can help researchers, students and educators, generate new knowledge about the socio-aesthetics of education and how the cultural politics of emotions (Ahmed, 2015) construct the communicative practices that inhibit relational awareness within and between individuals and groups (Damico et al., 2020; Kershaw & Nicholson, 2011; Mackey, 2016; Vettraino et al., 2017). Knowledge mobilization that shocks, incites or disorients without accounting for the emotional responses of an audience increases and prolongs defensive psychosocial structures. Nor is arousing fear and shame via didactic communication methods the most effective strategy for promoting changes in behavior (Curfman, 2009; Davenport, 2017; Lertzman, 2015). Alternatively, when one draws on monstrous feelings, events and spectacles through the arts, it opens inclusive spaces to identify and challenge systems of injustice (ecocide, environmental racism, colonization, sexism, heteronormativity). Mobilizing the power of storytelling, in the public sphere to critique social, political and structural inequities, illuminates how humans abuse their biomorphic and geomorphic powers while holding space open to generate inspired solutions for climate change dilemmas.

The dire impacts of the climate crisis will be inscribed on bodies. It will force those who currently enjoy a sense of comfort and security in the West to experience wild telluric vacillations between melancholia and ambivalence within a more aggressive sociality. While it is true the algorithms owned by white silicone-hearted power players have made the fight for climate justice exceedingly more difficult, I remain radically hopeful it is possible to employ creative transdisciplinary pedagogies across schools, universities and colleges to metabolize climate news while renewing who we are in relation to Others. In the field of education, we have a chance to create a massive shift in the climate narrative. To do this difficult work, creative educators from across the globe must challenge human narcissism and techno-optimism by inviting our monsters out to play.

References

Abraham, N. (1968). The shell and the kernel: The scope and originality of freudian psychoanalysis. In Abraham, N., & Torok, M. (1994). *The shell and the kernel: Renewals of psychoanalysis.* (pp. 79–106). (Vol. 1). (N. T. Rand, Trans.). University of Chicago.

Abraham, N. (1975). Notes on the phantom: A complement to Freud's metapsychology. In Abraham, N., & Torok, M. (1994). *The shell and the kernel: Renewals of psychoanalysis.* (pp. 169–179). (Vol. 1). (N. T. Rand, Trans.). Chicago, IL: University of Chicago.

Abraham, N., & Torok, M. (1971). The topography of reality: Sketching a metapsychology of secrets. In Abraham, N., & Torok, M. (1994). *The shell and the kernel: Renewals of psychoanalysis.* (pp. 157–164). (Vol. 1). (N. T. Rand, Trans.). University of Chicago.

Abraham, N., & Torok, M. (1972). Mourning or melancholia: Introjection versus incorporation. In Abraham, N., & Torok, M. (1994). *The shell and the kernel: Renewals of psychoanalysis.* (pp. 125–137). (Vol. 1). (N. T. Rand, Trans.). University of Chicago.

Ahmed, S. (2015). *The cultural politics of emotion.* Routledge.

Aljazeera (2020, July 28). Nearly 3 billion animals killed or displaced by Australian fires. (2020, July 28). *Aljazeera.* Retrieved from https://www.aljazeera.com/news/2020/7/28/nearly-3-billion-animals-killed-or-displaced-by-australia-fires'

Barone, T., & Eisner, E. W. (2012). *Arts-based research.* [Kindle DX version]. Retrieved from amazon.ca

Boal, A. (1979). *Theatre of the oppressed.* (C.A. & M.L. McBride, Trans.). New York, NY: Theatre Communications Group Inc. (Original work published 1974).

Boal, A. (1990). The cop in the head. *The Drama Review, 34*(3), 35–42.

Boal, A. (2002). *Games for actors and non-actors.* (A. Jackson, Trans.). Routledge. (Original work published in 1992).

Boal, A. (2006). *The aesthetics of the oppressed.* Routledge. Retrieved April 14, 2010, from http://lib.myilibrary.com.proxy1.lib.umanitoba.ca/browse/

Curfman, G. (2009). "Why it's hard to change unhealthy behavior – and why you should keep trying." Heartbeat: Harvard Health Publications. Available at www.health.harvard.edu/staying-healthy/why-its-hard-to-change-unhealthy-behavior

Damico, J. S., Baildon, M., & Panos, A. (2020). Climate justice literacy: Stories-we-live-by, ecolinguistics, and classroom practice. *Journal of Adolescent & Adult Literacy, 63*(6), 683–691. https://doi.org/10.1002/jaal.1051

Davenport, L. (2017). *Emotional resiliency in the era of climate change: A clinician's guide.* Jessica Kingsley Publishers.

Davis, C. (2010). The skeptical ghost: Alejandro Amenabar's the others and the return of the dead. In M. del Pilar Blanco & E. Peeren (Eds.), *Popular ghosts: The haunted spaces of everyday culture.* Bloomsbury. [Kindle Edition]. Retrieved from amazon.ca

del Pilar Blanco, M., & Peeren, E. (2010). Introduction. In M. del Pilar Blanco & E. Peeren (Eds.), *Popular ghosts: The haunted spaces of everyday culture.* Bloomsbury. [Kindle Edition]. Retrieved from amazon.ca

Derrida, J. (1994). *Specters of Marx: The state of the debt, the work of mourning and the new international.* New York: Routledge. [Kindle Edition]. Retrieved from amazon.ca

Eisner, E. (2010). Persistent tensions in arts-based research. In M. Cahnmann-Taylor & R. Siegesmund (Eds.), *Arts-based research in education: Foundations for practice.* (p. 2). Kindle DX version.]. Retrieved from amazon.ca (Original work published 2008)

Farrell, A. J. (2020). *Exploring the affective dimensions of educational leadership: Psychoanalytic and arts-based methods.* Routledge.

Fouché, J., Christiansen, C. T., & Lafrenière, M. J. et al. (2020). Canadian Permafrost stores large pools of ammonium and optically distinct dissolved organic matter. *Nature Communications, 11*(4500). https://doi.org/10.1038/s41467-020-18331-w

Freud, S. (1899). *The interpretation of dreams.* Basic Books. (2010).

Guattari, F. (2008). *The three ecologies.* (I. Pindar, & P. Sutton, Trans.). Continuum.

Jonze, S. (Director). (2013). *Her. [film].* Annapurna Pictures.

Kershaw, B., & Nicholson, H. (2011). *Research methods in theatre and performance.* Edinburgh University Press.

Kyle, K. (2020, September 16). *Researchers dig into Canadian North to understand carbon storage in permafrost.* CBC. https://www.cbc.ca/news/canada/north/carbon-storage-canadian-permafrost-research-1.5725436

Leavy, P. (2009). *Method meets art: Arts-based research practice.* The Guilford Press.

Lertzman, R. (2015). *Environmental melancholia: Psychoanalytic dimensions of engagement.* Routledge.

Mackey, S. (2016). Applied theatre and practice as research: Polyphonic conversations. *Research in Drama Education: The Journal of Applied Theatre and Performance, 21*(4), 478–491. doi: 10.1080/13569783.2016.1220250

McNiff, S. (2011). Artistic expressions as primary modes of inquiry. *British Journal of Guidance and Counselling, 39*(5), 385–396. doi: 10.1080/03069885.2011.621526

Morton, T. (2013). *Hyperobjects: philosophy and ecology after the end of the world.* University of Minnesota Press.

Nixon, R. (2011). *Slow violence and the environmentalism of the poor.* Harvard University Press.

NPR (2019, September 19). Transcript: Greta Thunberg's speech at the U. N. climate action summit. https://www.npr.org/2019/09/23/763452863/transcript-greta-thunbergs-speech-at-the-u-n-climate-action-summit

Perhach, P. (2017). Spotlight on tina yu. *Hi Fructose: The New Contemporary Art Magazine, 45.*

Reed, J., & Ehrlich, P. (2020). *My octopus teacher.* [documentary film]. Netflix.
Rothery, G. (Director). (2020). *Archive.* [film]. Independent; Hero Squared; Head Gear Films; Lipsync; Metrol Technology; Quickfire Films; Untapped; XYZ Films.
Smith, A. (2017). Bang sangho. *Hi Fructose: The New Contemporary Art Magazine, 45.*
Torok, M. (1968). The illness of mourning and the fantasy of the exquisite corpse. In Abraham, N., & Torok, M. (1994). *The shell and the kernel: Renewals of psychoanalysis.* (pp. 107–124). (Vol. 1). (N. T. Rand, Trans.). University of Chicago.
Vettraino, E., Linds, W., & Jindal-Snape, D. (2017). Embodied voices: Using applied theatre for co-creation with marginalized youth. *Emotional and Behavioral Difficulties, 22*(1), 79–95. doi: 10.1080/13632752.2017.1287348

10 Ecosophical Educational Research on the Edge of the Anthropocene

The Mental Health Commission of Canada (MHCC) (2012) reports that in any given year, one in five people lives with a mental illness. Mood and anxiety disorders are the most prevalent conditions for all ages (p. 9). Of those Canadians who have a mental illness, one million are children and adolescents between the ages of 9 and 19 years of age (p. 7). By 2041, the commission estimates the 12-month prevalence of mental illness in Canada will be 20.5%. Notably, there are relatively low rates of community service use of mental programmes. The MHCC report suggests the low rates of access are troubling when one considers children and youth who experience an untreated mental health issue are at much higher risk of experiencing mental illness as adults (p. 11). Given that many young people across the globe are experiencing more serious and frequent mental health issues, and the mental health systems for youth are often a hodgepodge of disjointed services, schools have been identified as important sites for mental health supports. What is problematic about MHCC's framing is the missing wider ecological context. We cannot be healed or studied apart from the ailing planet. If policymakers, healthcare workers and educators wish to provide more robust supports to young people who are emotionally injured by complex and compounding crises, research into the telluric emotions of the Anthropocene is a good space to begin a discussion about reorienting the field towards ecosophical ends.

The significance of educational research today must be, in some part, predicated on its ability to engage with the telluric emotions of the Anthropocene. Right now, there is paucity of work exploring the field's affective register. We cannot continue to use the same conceptual frames to study educational dilemmas that are situated in a field that is complex, integrative, multi-perspectival and affectively charged. To put it in blunt terms, educational research needs to get moody. Participants need to generate their own questions and work closely with researchers to continually examine the innovations and externalities that result from the research process. Research should promote and facilitate dialogue that fosters resilient engagement and courageous conversations. For these reasons, in this chapter I describe seven trajectories of ecosophical research and then propose nine areas of inquiry that focus on ecosophical aims and praxes, because we need new frames to describe *education as affect* in the wild and wilding sociality of the Anthropocene.

Seven Trajectories of Ecosophical Educational Research

With humility, I sketch seven trajectories of ecosophical educational research (EER), enmeshed and enlivened by the affective, social and ecological registers in my work. By definition, a trajectory is already on the move. I hope the concepts and provocations shared here will connect with the passion, wisdom and creativity of other educational researchers so we may explore new trajectories together. I would describe this as a process of *research echolocation*, an attempt to find and connect with Others who believe educational research is not limited to making eco-attunement possible, but can be itself the emergence of eco-attunement.

Trajectory 1: Ecosophical Educational Research Subverts Anthropocentric Frames

We are never outside of nature. Our mental and material ecologies are always intermingling within more-than-human ecologies. EER, then, confronts anthropocentric hubris and includes other parts of nature in its conceptualizations of subjectivity. It deconstructs the ways humans exert their biomorphic powers to deny the subjectivity of Others, and it resists anthropomorphic justifications of consuming beyond the Earth's carrying capacity. The Earth is undergoing a dramatic ecological transformation, and these changes threaten the continuation of life on Earth. EER recognizes the fraught imaginary boundaries between interior and exterior worlds as an integral research space to reconcile and take responsibility for one's positionality within the wider web of relations.

Trajectory 2: Ecosophical Educational Research Fosters a Deep Respect for Cultural and Biological Diversity

EER recognizes that monocultures deaden ecosystems and fields of research. A deep and abiding respect for the cultural and ecological diversity on Earth is essential. In this way, EER praxis resonates with *deep ecology* (Drengson & Inoue, 1996; Greenhalgh-Spencer, 2014; Naess, 2008), meaning its methodologies expand the number of people we identify with and fosters a sense of personal responsibility for Others. EER situates its analyses and subsequent judgements about education within a broader integrated ecological perspective to reflect the embeddedness of humans in complex ecosystems.

Indigenous, Black, Asian and other racialized people continue to be disproportionately impacted by the most severe impacts of global warming. Denial of the pain experienced by minoritized groups critically injures the soul of the planet. The escalation of white supremacist violence, xenophobia, misogyny and racism, all contribute to the loss of biodiversity, the scarring of landscapes, climate precarity and the increase in pandemics. EER demands we use multivocal approaches and the wisdom from all people to confront the oppressive social, political and economic structures fueling the ecological emergency.

To enliven a multiplicity of voices and perspectives, EERs use participatory visual arts research methods such as photovoice, digital storytelling and cellphilming *with* students and educators. Photovoice is a visual research method used to counteract injustices by sharing visual counternarratives that convey ways of seeing and relating in the world that have been traditionally marginalized (Spiegel et al., 2020). Participants express their point of view by taking photographs of scenes that relate to the collaborative inquiry. A digital story is a short, first-person digital narrative created by combining recorded voice, images, music or other sounds. A digital storyteller documents an experience, feeling or idea through a particular point of view using a digital narrative (Centre for Innovation in Education, 2021). Cellphilms are videos shot on a cell phone to explore an idea, a place, one's identity or one's relationships (MacEntee et al., 2016) to make a positive change in the world. Embracing the possibility that research can ripple across the multiple registers of life, the aims, questions, statements of significance and tentative conclusions of EER stretch forward and back in time.

Trajectory 3: Ecosophical Educational Research Turns Towards the Suffering Stranger

EER is intersubjective, in that it is inclusive of our more-than-human relatives. It is preoccupied with mutual recognition because it assumes our knowledge and our minds are deeply connected to the relationships we have with other people. EER troubles the self-sufficient and autonomous figure in education and it problematizes competitive frames and piecemeal epistemologies. EERs assert any attempt to justify the suffering of Others through mis-education, is morally obscene and they assume education plays a fundamental role in the turn-to-put-the-Other-first. EER studies how educators build vulnerability and responsibility for the Other into spaces of curricula and pedagogy and refocuses attention on the assumptions, behaviors and inherited practices that reify oppressive systems. The ethics of EER are reflected in one's capacity to see the suffering of Others and how recognition compels the compassionate witness to act. At its best, EER keeps the ethical imagination awake and mutual recognition as its goal.

Trajectory 4: Ecosophical Educational Research Tells Evocative Stories That Amplify Connectivity

The quality of the connections we have with Others are at the heart of understanding the emotional landscape of climate chaos. EERs often employ arts-based methods to expand an audience's visuality and to amplify their linkages among the more-than-human world. It is an art to communicate with others in ways that strengthen bonds while addressing a topic like climate change, a subject that often triggers tearful retreats, stunned silences or bursts of frustration. Hence, EERs' aim for evocative and interpretive spaces that can hold paradoxical feelings about phenomena that are deeply unsettling, even monstrous. We need a greater number of ecosophical "storytellers – a mighty force – to help us shift our mythology and imagine a future where together we thrive with nature" (Rodriguez, 2020, p. 121). EER situates its interpretive praxes in the more-than-human world to draw attention to the haunting beauty of its fragility and its life-giving force.

To tell compelling research stories about the affective dimensions of dis/connections to our more-than-human relatives, EERs work in partnership with young people and engage in processes, so the method and the content of research mitigates emotional distress and enhances connectivity. Making sculptures with natural materials within a do-no-harm ethos, photographing examples of mutualism and staging ethnodramas (Saldaña, 2008; Saldaña, 2010; Saldaña, 2011) outdoors so the liveliness of the landscape (birds singing, water gurgling, wind blowing) becomes a central character in the play are some examples of arts-informed EER methods. The EER research process, then, is a transitional space in which participants and researchers visualize how their unique lived experiences are reflected in the aesthetics of the more-than-human world.

Trajectory 5: Ecosophical Educational Research Makes Room for the Unbidden

EER acknowledges apocalyptic anxiety is an animating feature of the Anthropocene. In response to the uncertainty of the time, EER is a *making-with* the unbidden. It is responsive to the enigmatic unconscious in its use of methods to facilitate free association and its attention to drives and desires. EERs engage with the libidinal flow of research and they account for emotional composting in their methods. To open free associative spaces, they ask open-ended questions, follow the participants' dialogical threads, crawl below the surface of utterances and gestures, welcome divergent narratives and stay open during moments of discomfort that arise during the research process. Given the structuring power of the unconscious, EERs are ready for the latent content of formative relationships to appear unexpectedly in research encounters. They appreciate they can never fully know from where the participants speak, and that sometimes participants will speak more than they intend.

Trajectory 6: Ecosophical Educational Research Develops a Language of Telluric Emotions

Humans have no recent experience with the scale of environmental change we are experiencing right now. The speed and breadth of ecological degradation is eliciting waves of eco-anxiety through affective registers. EER develops a language in the field of education for telluric emotions because we do not live in an emotional vacuum, occupied by our own fear, grief, joy and ambivalence. Emotions are a socially contingent phenomenon. EER focuses on the relationships among changes in ecosystems, telluric emotions and the interior lives of individuals. For these reasons, EERs assume that throughout the research process, the field exerts an emotional force regardless of the status or asymmetry among the subjects. Research processes are always affected by the unknowable Other because our feelings get into each other's minds even before we have the words to describe we are having feelings.

To disavow something is to know it and to not know it at the same time. Disavowal can be more difficult to work with because it requires some aspects of reality to be accepted. It silences the critical voices in our heads and leads us to behave in ways which bring about the terrible situation we are trying so hard to escape. EER is concerned with the disconnection between the obligations inherent in loco parentis and the widespread evasion of climate precarity in spaces of curriculum and pedagogy. Consequently, EERs study the social organization of disavowal and learn from ambivalence, denial, enactments and other defence mechanisms people use to evade psychological pain in educative encounters.

Trajectory 7: Ecosophical Educational Research Moves Below the Surface

Relationship building with participants is critical. If possible, there should be multiple engagements with participants over a sustained period of time when engaging in EER. To move below surface talk, EERs listen carefully to their participants' stories and the stories the participants' unconscious defences, desires, silences and contradictions tell. Often, dialogic spaces open with a participant's reflection on a memento, artifact, song or photo to find an illuminating connection among the topic, the inner psychic landscape of the individual and the exterior world. As an illustration, EERs go for walks with participants to include more-than-human relatives in their studies.

The intersubjective approach in EER necessitates an artful attunement to the surroundings to render the complexity of the participants' lived experiences. Flexible and responsive, EERs allow competing understandings of the same phenomena to exist in the same research space. Further, EERs set the table for the ghosts of other meanings, relationships and places to infuse the interpretation of the participants' stories. EERs avoid premature interpretations of their participants' utterances, gestures and stories. They recognize how third-space unconscious processes and the researchers' own enactments can mute ideas, feelings or affects as a way to defend against what makes them uncomfortable in the research process.

Areas to Explore in Ecosophical Educational Research

It has become an ethico-political imperative for EERs to use their wisdom, privilege and passion to inquire into the frustration, anxiety, grief and hopelessness that haunt the landscapes of research and education in the Anthropocene. The last section of this chapter identifies nine broad areas (**A**) for educational researchers to consider if they are interested in pursuing intersubjective inquiries into the affective dimensions of teaching and learning amid significant ecological change.

A1: Cultivating Spaces of Mutual Recognition

Building a relational home for students is an integral part of processing difficult eco-emotions within educational encounters. We need to learn more about how to enhance the capacity of educators to cultivate spaces of mutual recognition and what types of dialogical praxes assist students in integrating strong feelings. This would involve a deeper exploration of *teacher as the withholding-witness* and *teacher as the compassionate-witness*. The intention would not be the creation of a new binary separating the two typologies, but an inquiry into the moments when teachers perceive themselves as shifting within and between the two dispositions in the same encounter. What causes one to shift to a withholding stance? What triggers more compassionate communication orientations in classrooms? These types of studies would not only assist researchers in identifying the characteristics of classrooms in which emotional nurturing and exchange are intentional, it would open space to think about the classroom as a transitional space in which young people can deal with their bewildering and disorienting feelings about a changing environment.

Resource depletion and climate volatility will cause greater civil unrest across the world. We need to learn more about telluric emotions and how communities of practice, in formal and informal educative spaces, foster mutual recognition in order to prevent disagreements from escalating into violence. Fighting and division among communities will become more prevalent due to climate change instability and young people will need a deep well of resources to draw upon in order to deal with interpersonal conflicts. Accordingly, we need researchers in education to illuminate and critique the communication strategies used by people in power to dismiss young people who are speaking out about the climate crisis, studies that interrogate dialogic processes used to disrupt binary thinking in emotionally charged encounters and inquiries into communication practices that foster interconnectedness and mutual understanding in face-to-face and online learning spaces.

A2: Educational Encounters of a Compassionate Kind

An agonizing experience in and of itself will not always exert a traumatizing effect. If there is sufficient attunement to a child's feelings by compassionate witnesses as the child's emotional pain is being integrated, the traumatic experience will be less likely to produce long-term effects. Part of a compassionate witness' work in maintaining encounters of a compassionate kind involves teaching young people to cultivate an inner climate of forgiveness amid emotional complexity. Weintrobe (2013) explains:

> Without an inner climate of forgiveness, in which we have real and not ideal expectations of others and ourselves, we may become stuck in a climate of hatred, bitter recrimination and relentlessness, easily feeling harshly judge, being harshly judging and not moving towards accepting the reality of the loss; we may feel caught in the ruinous expectation that our only salvation will be to have our ideal world back again, a world in which no loss has occurred.
> (p. 37)

It is actually the silence and the hypocrisies of adults that are some of the most traumatic aspects of emotional neglect (Blum, 2004). A Portkey for young people to develop an inner climate of forgiveness is to hear significant adults in their lives validate their feelings and apologize for the ecological mess we find ourselves in today. The lack of acknowledgement from adults is too difficult for young people to bear at this time in our history, *Okay, Boomer?* Acceptance of responsibility for the climate crisis involves speaking truthfully with young people about global warming, its causes and implications, without justifying the past in order to save face in the present. EERs could study the affective pedagogical dispositions of *dialogic fascination* in educative encounters, ones in which judgement, recrimination or dismissal exert tension.

A3: Creating Holding Environments

Unmetabolized anxiety can wreak havoc on our collective well-being. The more feelings can be made conscious, the more they can be used to foster processes of resiliency. It does not happen in isolation.

When people are affected by powerful feelings provoked by climate change, they need relational encounters in which those feelings can be contained. In an educational setting, containment could be provided by formal and informal networks of social supports or when teachers listen non-judgementally when students express shock, disappointment or anger about climate volatility. Listening to students' stories about climate change is the heart of mutual recognition and the foundation of holding environments for climate dread. EER focuses on how mutual recognition in the affective registers assists students in moving beyond their own defensive boundaries. ERRs also pay close attention to the ways educators model empathy building in emotionally charged spaces of curriculum and pedagogy. This important and complex work is an act of love, or what Patricia Hill Collins (2014) calls othermothering.

As an eco-feminist, I have often cringed when maternal rhetoric is operationalized to obfuscate structural inequities in education. *If you loved your students enough you would differentiate your instruction, pledge allegiance to the strategic plan, identify essential learning outcomes, get on board, stay afterschool to tutor students, stick to the script, buy school supplies for the class, go to your students' basketball games to build relationships, keep your head down and a steady supply of apples in your desk for students who never have enough to eat... and on and on and...* Maternal rhetoric has been used to frame women as bad objects or to place unreasonable expectations on women's bodies through paternalistic discourses of care. But despite some of the misuses of maternal rhetoric in the field, I believe it is instrumental in building an emotional holding environment to support the kind of social movement necessary to reorient schools towards ecosophical ends. For example, community mothering has been an essential aspect of civil rights struggles and a powerful organizing principle in Black communities (Westervelt, 2020, p. 250). As global warming intensifies, EERs may choose to explore the maternal aspects of teaching as an ethicomoral rallying cry, one that avoids tacitly shifting the burdens of climate change to students, while adults conveniently avoid the centrality of their own power to make a difference. *Teachers as community-mothers* are uniquely positioned to assist students in making sense of seemingly intractable and emotionally laden climate change issues, making this a worthwhile area of study:

We the Community Mothers in Education

We the Community Mothers in Education
see the Earth's ecosystems spin out of kilter
with wide eyes and clenched jaws
as our students' futures move into the Maybe column
We the Community Mothers in Education

know the tropes of anthropocentrism are shaped by race, class,
misogyny and colonialism
because we have read the herstory book
When the Land, Oceans, and Winged Ones are not treated
as ends unto themselves
the seeds of brutality take root
We the Community Mothers in Education
Use our voices and vocation to counteract the systems and practices
that frack the minds and potential of our students
Nurturing dialogic fascination with the other to
find more compassionate ways of living with other Beings
We the Community Mothers in Education
Teach our students to embrace the suffering stranger and to
do the math on our carbon fair share
We don't pathologize our students' feelings
As weavers of telluric emotions, we collectively mourn
species lost and the Earth's decreasing habitability
while refusing to abdicate our responsibility as
warrior culture makers
furiously writing counternarratives
with our pedagogy
to create a livable and
loving future
for all

A4: Attunement in Educative Spaces

The pioneering work of Harold Searles (1972), Leslie Davenport (2017), Renee Lertzman (2015), Caroline Hickman (2019; 2020) and Sally Weintrobe (2020) highlight the central importance of emotional attunement in emotionally charged discussions about environmental degradation. Attunement refers to asking others how they are feeling, giving language to an emotional reaction, self-analysis about one's own emotional responses and treating the Other's emotional utterances from a place of compassion. We need to better understand what gets in the way of attunement and what it looks like to name one's emotional experience, foster self-compassion and talk about the ethico-political relations enmeshed in the mis/recognition of climate affects in public education. Furthermore, the field would be enriched by explorations of the relationships between students' feelings and struggles and the systems and structures perpetuating the anthropogenic causes of climate volatility. As an example, studies of counselling services in schools need to account for telluric emotions and critiques of the therapeutic practices that lead to pathologizing individual students' emotional stressors within a medical model. Research into the eco-relational attunements in the guidance counselor's office are urgently needed as students prepare to graduate in a world that is becoming less livable.

A5: Interdependence: Emotional Learning Within a Wider Web of Relations

There is a paucity of research in the field of education that focuses on pedagogy oriented around learning in emotional harmony with the natural world. Ecosophical researchers could explore connections between learning *from* the more-than-human world and the development of emotional resiliency processes. We know learning outside fosters a sense of personal well-being among children (Coates & Pimlott-Wilson, 2019; Malberg Dyg & Wistoft, 2018; Waite, 2010; Waite, 2011), but we need to learn more about how the emotional connections young people develop to the Land, Water and Sky infuse the emotional currents that shape their understandings of themselves in relation to other Beings. We could draw from the work of scholars like Catherine Kelly (2021), who studies psychological responses to *blue spaces* (water environments) and how people can integrate blue mind practices to enhance their individual and collective well-being. I am currently designing a study to explore students' emotional connections to the vast prairie sky and its connections to the air they breathe.

There are pockets of teachers and students who are engaged in transdisciplinary inquiries about the impact climate crisis is having in their own communities. If researchers brought these educators and students together in eco-learning networks, we could discover valuable insights about how eco-networks assist or inhibit different groups' abilities to process eco-anxiety. There would also be much to learn about the emotional labour involved in collectively imagining new images of the future in which we protect, honour and find joy in the awe-inspiring natural world. How might networks of ecosophical teachers and students foster emotional resiliency processes as they internalize the impacts of ecological degradation and work in solidarity to mitigate the damage?

A6: Dialogic Fascination as a Cure for Pathologizing Ugly Emotions

Critics worry a turn to the mental health profession to inquire into the affective dimensions of the climate crisis will inevitably birth a new clinic (Foucault, 1994) staffed by eco-conscious professionals who will find, diagnose and treat a new cohort of patients within a medicalized model. A lecture Foucault (2014) gave on April 2, 1981, offers some important cautions about neologisms like *eco-anxiety* and *climate grief* becoming more common in the public discourse. He begins the lecture with a disturbing account from 1840 on "the moral treatment of madness" (p. 11). He recounts several treatment sessions between a psychiatrist Leuret and his patient. Doctor Leuret asks his patient to admit what he sees and hears is evidence of madness. The patient resists and repeatedly states he is not mad. After each denial, Leuret drenches his naked patient with cold water. The cold water "treatment"' does not stop until the patient confesses he is insane. From the story, Foucault surmises the "patient" must place himself "in a relationship of dependence with regard to another, and modify at the same time, his relationship to himself" (p. 17). In this sense, social pathologies are attributed to the deficits in individual actors. With regards to the language of climate change emotions, the relevant danger lies in labelling a student as a sufferer of eco-anxiety inside a frame which hides the root causes of anthropogenic climate change in a medicalized model. Further, there is a danger of mis-characterizing healthy responses to climate anxiety in service of entering novel sets of criteria in future editions of the *Diagnostic and Statistical Manual of Mental Disorders* (American Psychiatric Association, 2013). EERs could keep Foucault's warning close as they engage in research about the affective dimensions of climate change in the field of education so we do not birth an eco-clinic.

A7: Eco-Anxiety

The epoch of the Anthropocene is characterized by generalized state of eco-anxiety. We need to learn more about how eco-anxiety manifests in classrooms and how teachers can create holding environments for the existential angst of their students. Lertzman (2020) identifies four tendencies listeners exhibit when hearing others express anxiety. They include righting; educating; cheerleading; and guiding. She says when someone is *righting*, the person lets people know that if they truly cared about climate change they would change their behavior. *Educating* refers to waking people up with compelling information so they can act differently after they learn new information. *Cheerleading* is to try and overwhelm people with positivity and flattery so they continue to listen to you. Finally, *guiding* is listening to someone and then listening some more before you offer information or advice. Lertzman (2020) suggests *guiding* is significantly more effective when trying to learn with people in times of significant change. Research that accounts for the emotional dimensions of climate change could illuminate how guiding processes in classrooms alleviate the embodied climate stress students and educators internalize and project in educative spaces.

A8: Trauma Stewardship

Teaching and leading during COVID-19 has made it clear there is a gap in the literature about trauma stewardship in the field of education. Trauma stewardship refers to the way people in the helping professions come to do their work, how they are affected by it and how people make meaning about their work (van Dernoot Lipsky & Burk, 2009). As the climate crisis intensifies, we need robust research programmes to examine how educators are bearing witness to the climate crisis, how they are looking after their own well-being and how the affective register shifts inside other registers as a result of the cascading climate crisis. Resources will need to be allocated to provide acute mental health supports in school communities hit by climate disasters and we will need researchers who will sensitively engage in relational assessments about the effectiveness of those acute supports in school communities. In times of crisis, teachers may be called upon more often to support students and their families and we need to understand the psychological impact eco-trauma stewardship has on the ability of teachers to act as compassionate witnesses.

A9: Ecosophical Approaches to Change

Peter Senge once said, "People don't resist change, they resist being changed". We need EER to develop a psychoanalytically informed language to help educators navigate the currents of messy emotions arising during school change processes. For example, how can teachers increase their tolerance of intense feelings and disturbing situations precipitated by climate change? How might they validate and normalize the disappointment, anger and sadness of Others when global warming forces school systems to shift priorities and reallocate resources? We know people are adept at creating distance between themselves and pain, so we will need dialogical spaces in service of understanding the productivity of resistance and how to attune to the emotional life of staff members who express ambivalence about broad-scale change. Correspondingly, we need to learn more about the shifts leaders make between their stances as withholding and compassionate witness in times of crisis.

As we become more conscious that educators are preparing young people to live in a world that is prepared to go on without them, educational research, that focuses on individual attainment and medical models of mental health, is out of time. What is called for now are spaces of EER, to move among the affective ecologies entangling subjectivity in the Anthropocene. Valuing an exploration of the dis/connections among the inner and exterior lifeworld of oneself and Others is a place to begin, so love, forgiveness, anxiety, grief and hope are brought to bear in conversations about the complexity of what it means to become an educated person in the Anthropocene. We face an existential crisis, and the old narratives rooted in dominance are going to take us to the brink. The promise of educational research is that we can weave new stories. Instead of amplifying narratives that valorize competition and individual achievement, we could weave stories in education that define "success" in relation to responding to the suffering of Others, mutual recognition and eco-justice.

References

American Psychiatric Association. (2013). *Diagnostic and statistical manual of mental disorders* (5th ed.). https://doi.org/10.1176/appi.books.9780890425596

Blum, H. P. (2004). The wise baby and the wild analyst. *Psychoanalytic Psychology, 21*(1). 3–15. doi: 10.1037/0736-9735.21.1.3

Centre for Innovation in Education. (2021, April 18). *Digital storytelling.* https://www.liverpool.ac.uk/centre-for-innovation-in-education/resources/all-resources/digital-storytelling.html

Coates, J. K., & Pimlott-Wilson, H. (2019). Learning while playing: Children's forest school experiences in the UK. *British Educational Research Journal, 45*(1), 21–40. https://doi.org/10.1002/berj.3491

Davenport, L. (2017). *Emotional resiliency in the era of climate change: A clinician's guide.* Jessica Kingsley Publishers.

Drengson, A., & Inoue, Y. (Eds.). (1996). *The deep ecology movement: An introductory anthology.* North Atlantic Books.

Foucault, M. (1994). *Birth of the clinic: The archaeology of medical perception.* Vintage.

Foucault, M. (2014). *Wrong-doing truth-telling: The function of avowal in justice.* F. Brion & B.E. Harcourt (Eds.), S.W. Sawyer (Trans.). The University of Chicago Press.

Greenhalgh-Spencer, H. (2014). Guattari's ecosophy and implications for pedagogy. *Journal of Philosophy of Education, 48*(2), 323–338. https://doi.org/10.1111/1467-9752.12060

Hickman, C. (2019). Children and climate change: Exploring children's feelings about climate change using free association narrative interview methodology. In P. Hoggett (Ed.), *Climate psychology: On indifference and disaster.* (pp. 41–59). Palgrave Macmillan.

Hickman, C. (2020). We need to (find a way to) talk about … eco-anxiety. *Journal of Social Work Practice, 34*(4), 411–424. https://doi.org/10.1080/02650533.2020.1844166

Collins, H., P. (2014). *Reconceiving motherhood.* K. A. Story (Ed.). Demeter.

Kelly, C. (2021). *Blue spaces: How and why water can make you feel better.* Welbeck Balance.

Lertzman, R. (2015). *Environmental melancholia: Psychoanalytic dimensions of engagement.* Routledge.

Lertzman, R. (2020). *From Anxiety to Action: How to Stay Sane While Fighting Climate Change* [Webinar]. PSYCHALIVE. https://www.psychalive.org/pl_resources/28337/

MacEntee, K., Burkholder, C., & Schwab-Cartas, J. (2016). *What's a cellphilm: Integrating Mobile phone technology into participatory visual research and activism.* Sense Publishers.

Malberg Dyg, P., & Wistoft, K. (2018). Wellbeing in school gardens: The case of the gardens for bellies food and environmental education program. *Environmental Education Research, 24*(8), 1177–1191.

Mental Health Commission of Canada. (2012). *Making the case for investing in mental health in Canada.* https://www.mentalhealthcommission.ca/sites/default/files/Investing_in_Mental_Health_FINAL_Version_ENG_0.pdf

Naess, A. (2008). *The ecology of wisdom: Writings by arne naess*. A. Drengson & B. Devall (Eds.). Counterpoint.

Rodriguez, F. (2020). Harnessing cultural power. In A. E. Johnson & K. K. Wilkinson (Eds.), *All we can save: Truth, courage, and solutions for the climate crisis*. (pp. 121–127). One World.

Saldaña, J. (2008). Ethnodrama and ethnotheatre. In J. G. Knowles & A. L. Cole (Eds.), *Handbook of the arts in qualitative research: Perspectives, methodologies, examples and issues*. (pp. 195–207). Sage.

Saldaña, J. (2010). The drama and poetry of qualitative method. In M. Cahnmann-Taylor & R. Siegesmund (Eds.), *Arts-based research in education: Foundations for practice*. (p. 2). Kindle DX version.]. Retrieved from amazon.ca (Original work published 2008)

Saldaña, J. (2011). *Ethnotheatre: Research from page to stage*. Routledge.

Searles, H. F. (1972). Unconscious processes in relation to the environmental crisis. *The Psychoanalytic Review, 59*(3), 361-374

Spiegel, S. J., Thomas, S., O'Neill, K., Brondgeest, C., Thomas, J., Beltran, J., Hunt, T., & Yassi, A. (2020). Visual storytelling, intergenerational environmental justice and indigenous sovereignty: Exploring images and stories amid a contested oil pipeline project. *International Journal of Environmental Research and Public Health, 17*(7), 2362. https://doi.org/10.3390/ijerph17072362

van Dernoot Lipsky, L., & Burk, C. (2009). *Trauma stewardship: An everyday guide to caring for self while caring for others*. Bernett-Koehler Publishers.

Waite, S. (2010). Losing our way? The downward path for outdoor learning for children aged 2-11 years. *Journal of Adventure Education and Outdoor Learning, 10*(2), 111–126. https://doi.org/10.1080/14729679.2010.531087

Waite, S. (2011). Teaching and learning outside the classroom: Personal values, alternative pedagogies and standards. *Education 3-13, 39*(1), 65–82. https://doi.org/10.1080/03004270903206141

Weintrobe, S. (2013) The difficult problem of anxiety in thinking about climate change. In S. Weintrobe (Ed.), *Engaging with climate change: Psychoanalytic and interdisciplinary perspectives*. (pp. 33–47). Routledge.

Weintrobe, S. (2020). The difficult problem of anxiety in thinking about climate change. In S. Weintrobe (Ed.), *Engaging with climate change: Psychoanalytic and interdisciplinary perspectives*. (pp. 33–47). Routledge.

Westervelt, A. (2020). Mothering in an age of extinction. In A. E. Johnson & K. K. Wilkinson (Eds.), *All we can save: Truth, courage, and solutions for the climate crisis*. (pp. 249–254). One World.

11 Ecosophical Educational Leadership

The work of educational leaders is often propelled by high stakes accountability frameworks within an entrepreneurial culture. Large parts of the field have devolved into the datafication of schooling, in which training the next generation of workers has become the proxy for social justice. Relentlessly driven to close achievement gaps in performative systems, the moral purpose of school leadership is subsumed by elaborate fantasies that mask the incongruencies and problematic intersections among the democratic aims of education and social mobility. Mechanisms, like standardized tests, frame schooling as competitive system, in which external accountability measures are instruments of the social good. Under the guise of inclusion, neoliberal policies are the wolf in sheep's clothing, compelling leaders to internalize feelings of guilt and shame when bad test scores are strategically severed from the wider social context.

In neoliberal frames, learning is associated with economic success and the acquisition of material goods. It dissuades students and teachers from critiquing the same economic system which is responsible for ravaging the planet. As the climate crisis intensifies, educational leaders will need to work through their own ambivalence and reconnect with the moral purpose of leading. Ecosophical educational leadership research (EELR) can play a significant role in developing images of practice that focus on ethical leadership encounters inclusive of our more-than-human relatives. In this chapter, I build on the seven trajectories of EER, discussed in the last chapter, and identify some areas of EELR grounded in radical hope and problematizing asymmetrical communication processes that reify entrepreneurial frames of leadership.

I define leadership as a set of unfolding affective relations in the field that emerge organically in the interactions between Others as they pursue intersecting and competing interests (Farrell, 2020). The definition assumes I am changed by my experiences, but my experiences also change me. With this in mind, I want to begin with the power of the example, a trajectory moving from a leader's interiority to the exterior field of relations, as an area of EELR. There are school and district leaders whose lived experiences are images of praxis. These are people who ignite students' natural curiosities about the more-than-human world, open dialogic spaces about significant changes to the Earth's ecosystems, build outdoor classrooms and encourage teachers to facilitate land-based learning in spaces of curriculum and pedagogy. Researchers could amplify the stories of ecosophical educational leaders (EELs) who lead with a sense of wonder and an appreciation of their place within the natural world, and drawing the stories together, could make a positive impact in a field dominated by structural-functional frames of educational leadership.

If educational leaders commit to more ecosophical ways of being in their lives, they will often find themselves working with difficult knowledge and advocating for praxis that grates against entrepreneurial frames. Due to the complexity of ecosophical leadership work, environmental psychologist and economist Per Espen Stoknes (2015) suggests five different "flips" leaders should consider to minimize the operationalization of psychological defences against difficult knowledge. His first suggestion is to flip what is distant to social. Make an issue near, personal and urgent and share social norms that are positive to enhance peer to peer influence. The second suggestion is to flip from doom to supportive. Leaders should connect environmental stewardship to human health and flourishing. A third suggestion is to flip dissonance to simpler actions by making climate friendly behaviours the default or more convenient. He uses the example of using smaller plates at a buffet restaurant. When people use smaller plates, they waste less food. A fourth recommendation is to flip denial to tailoring signals to visualize progress and providing motivating feedback to people. Finally, he encourages leaders to flip tribalism with better stories. He asserts that leaders need to tell stories about where we all want to go and amplify the work of people who are already making positive changes. EELR could contribute to this work, by identifying the types of psychological defences at work in educative encounters and amplifying educational leadership praxis that addresses psychological defences within a relational home.

Relational Home Maintenance

For ten years, I worked as an educational consultant for a non-profit organization in Manitoba, Canada. My work involved supporting school leaders who were embarking on significant social justice initiatives in their schools. During that time, I kept detailed journals to remember the important things I learned from school and systems leaders. Slivers of memories are tightly packed between the lined pages across several notebooks. Rereading the contents of the journals has been incredibly valuable to me as an instructor and researcher who is focused on the affective dimensions of change. To think about the relational aspects of leadership in precarious times for this text, I turned to my personal archive. Although I read new things into the pages each time I revisit them, sacrifice has remained an important dialogic motif, particularly in my descriptions of encounters with senior administrators. Like accountants of lost time, many of them kept track of how many evenings they spent emailing, the amount of family dinners missed and the obscene number of hours they spent in windowless offices in what one school principal referred to as "machine mode". Leaders in perpetual machine mode were often praised when they skipped lunch breaks and they secured promotions to lead bigger schools by beginning and ending their work days in the dark. Sacrificing time with family, friends and connections to the natural world continues to be normalized as expectations for successful school and divisional administrators. EELR could explore shifts in the affective dimensions of "machine-mode" and examine the relationship to between machine-mode and one's feelings of eco-estrangement or connections to place.

When leaders get into machine-mode, they can be tempted to look for expedient relational trajectories to address external demands when daily tasks exceed the time they have during the work week (Winton & Pollock, 2016). When superintendents or policymakers foist new initiatives on schools, principals have to generate additional energy and goodwill in the affective register. In response, the relationships among colleagues reorient towards the burden of the external demand, the needs of the formal leader and away from life-enhancing relational bonds. Both the imposition of an external mandate, along with the increases in stress during times of change, fuses relationship building to instrumental ends. A page in one of my journals elicits an emotional pinch about the cultivation of relationships as a leadership strategy. In a conversation with a seasoned school principal, he asserted, "Continuous school improvement requires teachers to allocate time to school-wide-priorities. Time is precious, they already work hard. If they like you and what you're doing is meaningful to them, they'll give you the time. Relationships are everything. Make it meaningful".

Sometimes a school improvement process produces more of itself, in order to continue to be itself. It is under its own power but demands constant servicing and a greater presence in the lives it seeks to improve. The maintenance of strategic plans has become an unquestioned leadership practice (Marshall & Sogaard Nielsen, 2020), but as the daily work of educators expands and intensifies, there is less energy available to run on the school improvement treadmill. When teachers and leaders cannot bridge the everyday divide between the fantasy of how they would work if they had the resources to function to the best of their ability, and the reality of how they can work given the myriad of hurdles in their way, they engage in service rationing (Lipsky, 2010). Service rationing can be an effective temporary coping mechanism to deal with the effects of machine-mode, but it often strains relationships when people elect to suspend tasks or give up responsibilities impacting others. EELR could explore the prevalence of complementary structures in times of service rationing and how complementary structures contour a leader's ability to build a relational home for Others.

In a complementary structure, dependency becomes coercive (Benjamin, 2018). These are the moments when people demand recognition for themselves and expect Others to go without. In other words, each person becomes increasingly concerned with being right and leveraging power rather than constructing a shared perception of reality. One is forced to choose between submission or resistance to the Other's demands (Ogden, 1994). It is why I bristle when my eyes linger over the words "Make it meaningful" and "they'll give you the time" in my journal entry, because the words speak to the entanglements among complementary structures, relationship building and the extraction of resources from people in organizations. Like exhausted affinity conjurers, leaders try and spark life into wilting school improvement initiatives by placing a high value on emotional management to achieve the ends of the organization. Periods in which the goals of the organization supersede mutual recognition and the moments a leader becomes a disenchanted witness would be worthy areas of EELR.

Over the years, I have heard many leaders make the case against being visibly vulnerable in times of crisis. The persona of the unflappable heroic leader remains a troubling trope because it deludes people into thinking they can enter the pain of Others when they are detached from their own pain. The most authentic way to build a relational home for Others, which I understand as an ethical imperative of educational leadership, is to cultivate the ability to observe one's feelings and the feelings of Others with curiosity. Critical studies of the socialization forces that influence educational leaders' perceptions of what emotions are permissible to express publicly in challenging times have come from feminist (Blackmore, 2006; Blackmore, 2013; Khalil & DeCuir, 2018; Marshall & Young, 2013) and queer (Lugg, 2003; Lugg, 2017; Tooms et al., 2010) scholars. EELR could add to this important body of work by examining the relationship among inherited leadership myths produced in entrepreneurial school cultures, a leader's somatic awareness and the leakiness of holding environments in educative spaces.

Holding environments are more porous in times of heightened stress, which makes containing anxiety, grief and catastrophic thinking an area of concern for EELRs. Educational leaders are called upon to provide support to students, families and teachers in times of crisis. Different types of crises (personal, cultural, ecological) vary in size, level of urgency and duration, each one producing distinct challenges. A crisis can inflict a major shock, undermining or limiting the very capacities a community needs to deal with a crisis. Part of the work of crisis leadership, then, is to contain the depth and duration of the chaos, bewilderment, helplessness and anger it precipitates, and to mobilize and harness the coping capacity from within the community (Boin et al., 2017).

Ecosophical Educational Leadership in Times of Crisis

One thing we can be certain of is that as the area of land capable of supporting human life shrinks, a traumatizing vulnerability will be exposed. The precarity of human existence will haunt the absolutisms of life in the light of day. When no continuity of existence can be assured, more people will conform to the abusive ethos of capitalism, then *we are here for a good time, not a long time, so let the consuming roll!* At the same time, safety and protection will become superordinate in state responses to social tumult, and as history has taught us, these re/actions will lean more towards brutal force and less in the direction of self-sacrifice, radical sharing and compassion, unless we make a conscious effort to redirect our attention and interests.

Trauma has existential and intersubjective dimensions. As climate precarity further shatters our metaphysical illusions in the field of education, the roll of emotional dwelling with Others will become an integral area of EELR study and praxis. How will educational leaders enter into the Other's pain without retreating from the traumatic state? What is the personal and collective threshold in educative spaces for confronting eco-anxiety and grief head on? Educators and students will have to feel their way through educative encounters in answer to those questions. EELR, may be able to open dialogical spaces to tackle the monstrous feelings precipitated as we confront the possibility that Earth as a dwelling place is going to disappear.

In the Anthropocene, an aim of education must be to create a relational home for a traumatized state of Being. In the absence of the eco-analyst figure, educational leaders and their community can become one together. To assist Others in facing ecological trauma head on, educational leaders must be able to empathize with Others while being stressed within the same affective register, strive for mutual recognition even when it cannot be reciprocated and move beyond their own psychological defences. EELRs can play a pivotal role in reconceptualizing educational leadership as process to make telluric emotions more bearable as leaders learn how to engage in this complex and affectively charged work.

Trauma Exposure Responses as an Area of Research in Educational Leadership

As the ecological register decays, so does the affective one. Unmetabolized anxiety grows as an invasive species in the field despite our best attempts to insulate ourselves in happy affects. Subconsciously, losses of biological diversity are introjected, precluding an ability to mourn. Due to a socializing disavowal of the seriousness of our situation, the inability to mourn real and anticipated losses in biodiversity produces a traumatizing effect (Abraham & Torok, 1972). When viewed through the lens of traumatology, the climate crisis exerts a higher order trauma (Woodbury, 2019) because it is ubiquitous. It affects all aspects of our lives and enlivens the ghosts of past traumas because it is a phenomena humanity has yet to collectively acknowledge.

EERs are concerned with the trauma exposure responses (TERs) in relation to an educational leader's in/ability to build a relational home for staff and students. There is a paucity of research in the field of education that addresses how to recognize TERs in oneself and in Others. Although it is true TERs manifest differently in different people, there are some patterns that emerge when people in the helping professions are witness to the suffering of Others over long periods of time. Some of the TERs exhibited in educational leaders and teachers as the climate crisis intensifies will include anger and cynicism. I recall a student in one of my fall classes getting very upset with me in a Zoom meeting because I invited the class to read an article about the relationship between the COVID-19 pandemic and the climate crisis. He rolled his eyes and then yelled, "God, isn't it enough to think about leading in a pandemic, now you expect us to add climate change to the list!" When we reach the limit of what we can bear, *anger* can mask one's pain and anxiety.

To protect ourselves from the daily deluge of emotional intensity, we often externalize defence mechanisms to protect ourselves. This can manifest as the *inability to listen to others* or in *reductions to dialogic complexity*. Nuance disappears and staking out positions begins. When people engage in black and white thinking, it is because their mind has become rigid. They are less improvisational in their responses and they find it increasingly difficult to consider the views of others. People might become highly reactive, especially when they *feel persecuted*, which becomes a breeding ground for singlemindedness and expressing certainty about the supremacy of one's perspective. EELR has a role to play in studying the emergence and influence of repetitions in educational leadership encounters that amplify and add layers of uncertainty, self-judgement and the filtering out of positive aspects of the Others' perspective in dialogical spaces.

A TER common among educational leaders is *I can never do enough*. The feeling of coming up short over an elongated period of time can lead to another TER, *guilt*. Some of the school leaders I spoke to amid the COVID-19 outbreak said they felt extremely guilty about not working hard enough or not being able to effectively support teachers who struggled with remote teaching. Unfortunately, not doing enough often elicits thoughts about *not being enough*. One principal revealed, "I don't know if I'm having any kind of an impact anymore", and a vice-principal disclosed, "Maybe this work isn't for me. I'm losing myself in all of this".

By making trauma exposure communicative, the defences enacted to hide the source of the real pain paradoxically reveal the psychosocial structures and the source of the fear and pain. Bromberg (2011) describes it as:

> A dissociative mental structure is designed to prevent cognitive representation of what may be too much for the mind to bear, but it also has the effect of enabling dissociatively enacted communication of the unsymbolized affective experience. Through enactment, the dissociative affective experience is communicated from within a shared "not-me" cocoon.
>
> (p. 21)

EERs could make a significant contribution to the field by engaging educational leaders in reflective dialogue about instances when the "not-me cocoon" enlivens defence mechanisms in times of significant change and how unnamed feelings in leadership encounters become bad house guests in the relational home with Others.

Relational homes are open to a diverse range of intersecting and competing viewpoints and lived experiences. EELs care for the relational home by privileging mutuality, reflecting on their own life story and emotional tapestries, adopting a stance of curiosity before they pass judgement and creating affirming stories with others. Some of the stories EERs would find compelling, as they relate to TERs, are uchronic tales (Portelli, 1988). Uchronic stories are the narratives of what should have happened, but did not. When these narratives are woven in dialogical spaces, they provide key insights into the desires, fears, fantasies and disappointments of the weavers. The conflicting desires of the antagonists and protagonists juxtaposed within conflict frames can illuminate the un/productive ways people interpret and respond to mounting losses. When Others find their own meaning in uchronic stories, they may be better able to assess how acting as a compassionate witness can alter the field of relations from the inside out.

Educational Leadership as a Sympoietic Pulse

In *Staying with the Trouble*, Donna Haraway (2016) describes sympoiesis as a *making-with*. She asserts, "Nothing makes itself; nothing is really autopoietic or self-organizing...Sympoiesis is a word proper to complex, dynamic, responsive, situated, historical systems. It is a word for worlding-with, in company. Sympoiesis enfolds autopoiesis and generatively unfurls and extends it" (p. 58). EERs understand educational leadership as sympoiesis, a making-with that extends beyond the temporal and spatial boundaries of schools. Thinking about leadership as sympoeisis, EER troubles epistemologies grown in competition, hyper-masculinity and rugged individualism and it problematizes descriptions of leadership that distill leading into the actions of individual people. Leadership as sympoeisis exists amid a field of relations inclusive of our more-than-human relatives.

Motivational Interviewing as a Sympoietic Pulse in Educational Leadership

Climate scientists have been warning us for years about the impacts of global warming, yet the work of educational leaders does not convey the urgency of our collective situation. The fifth IPCC Special Report (2018a) assessed the projected impacts of global warming at the level of 1.5 degrees Celsius above pre-industrial levels. The IPCC projects, with a high level of confidence, warming beyond 1.5 degrees will cause additional long-term changes in the climate system such as sea level rise, species loss and increases in ocean acidity (IPCC, 2018b). In another study, Warren et al. (2018) project that with current pledges to reduce greenhouse gas emissions, global warming will breach 3 degrees Celsius above pre-industrial levels. This will result in climatically determined geographic range losses of greater than 50% for 49% of insects, 44% of plants and 26% of vertebrates. At a warming level of 1.5 degrees Celsius above pre-industrial levels, climatically determined geographic range losses fall to 6% of insects, 8% of plants and 4% of vertebrates. There is still time to act to mitigate some of the worst effects of global warming but it means we are going to have to expand the field's visuality to include the climate breakdown and open dialogical spaces to help each other mitigate impacts within the ecological, affective and cultural register in our lives.

In spite of the deluge of information available at the click of a computer mouse, most leaders in countries that have suffered the least from the impacts of global warming remain ambivalent about their role and the role of education in mitigating the damage. Stuck in a cult of presentism, educational leaders, teachers, politicians and the wider public evade their responsibilities for the lives of generations of school children yet to come. If we are to limit global warming to 1.5 degrees Celsius above pre-industrial levels, EER has a role to play in wrestling with climate change ambivalence, particularly in the countries that are most responsible for global emissions. We know in times of dramatic change people hold conflicting attitudes. They find reasons to change (*I'm worried about rising sea levels. Higher temperatures in the prairie regions of Canada and the US are diminishing bird habitats. Asthma rates are off the charts now.*) and they look for reasons to stay the same (*it will kill the economy to get off of fossil fuels. Other countries are not making the necessary changes, so why should we take the hit? The problem is too big. It's already too late*).

To evade the suffering of Others impressing itself on one's self-satisfied life, one frequently becomes resistant to persuasion and defensive against blame. A method that leaders can use to elicit productive change talk is motivational interviewing (MI). MI emerged in the field of addiction treatment. Miller (Miller & Rollnick, 2013) conceived the approach while counselling people who were struggling with alcoholism. Since then, MI praxis has evolved and the method is used in other fields. The essence of MI is to support change in a manner that is congruent with a person's own values and concerns. It is a dialogical method to guide people in resolving their own inner ambivalence (Miller & Rollnick, 2009; Rollnick & Miller, 1995). Its method runs counter to the many ways educational leaders are socialized to solve problems for other people. In the name of caring for other people's children, they debate, cajole, argue, flatter and coach to get other people to change. However, the process of MI encourages leaders to resist their compulsion to solve problems for other people, a compulsion referred to as the *righting reflex* (Miller & Rollnick, 2013) in the literature. The goal of MI is to create a collaborative space so the person you are guiding in the conversation ends up debating with themselves. The process is intersubjective, in that the person who facilitates listens intently to the Other in order to guide *and* follow in conversation.

Typically, the MI method is described in four steps. It is important to note the "steps" can be disorderly and repeated, making it a process and not a recipe. The intent is to create a third space (Benjamin, 2018) in which the utterances and gestures of those in the encounter create something novel outside of both people. The first step of the process is to build a strong relationship. It is difficult to share what matters most to us if we do not believe the Other will take care of us and what we have to say. The next part of the process involves settling into a focus for the conversation. In this step, tensions with an intersubjective psychoanalytic approach emerge. Unlike the free associative space in the analyst's consulting room, the MI guide tries to keep the dyad working on the same topic. In the third phase, there is an attempt to evoke change talk by asking open-ended questions, giving affirmations, mirroring what you think you heard the other person say and then summarizing the conversation. In the last stage, the partners sketch a plan to help the person achieve what he or she sets out to do. MI is not a silver bullet. It should never be used in disciplinary conversations, nor is it a series of techniques to get people in organizations to act in the leader's best interests. Rather, it is an image of practice one can use to facilitate emotionally laden decision-making processes. EERs may find the MI process a worthy methodology or a supportive frame to practice one's relational interviewing skills.

Ecolinguistics and Ecosophical Educational Leadership Research

Beginning with tongue in cheek, I often allocate time at the front end of my educational leadership grad classes to discuss the *dark arts of communication* wielded by people in power, who leverage emotionally charged words, phrases or stories to satisfy their own interests. I want students to become more conscious of the communication strategies pulling society further away from life-affirming orientations, no matter what side of the aisle the words emanate from. I have been inspired by the work of ecolinguists who study how language shapes our affective, cultural and ecological registers. The "...*eco* of ecolinguistics refers to the life-sustaining interactions of humans with other humans, other organisms and the physical environment, with some kind of normative orientation to protecting the flourishing of life" (Stibbe, 2021, p. 9). EELR, which draws from the field of ecolinguistics, is deeply concerned with the words and texts surrounding educational "leadership" and their relationship to thriving within the Earth's ecological limits.

In a recent grad class, my students deconstructed the support documents released by the Manitoba government on Bill 64, The Education Modernization Act. Several problematic communication strategies were found in the government's *Better Education Starts Today: Putting Students First* document (Manitoba Government, 2021). Empty signifiers such as *accountability, oversight, modernization, consistent, efficient* and *increase teacher and leader effectiveness* were wrapped in an entrepreneurial frame to reify the logics of neoliberalism. Although the government tried to soften the report in places by claiming to *put students first*, the first pillar of the report is called "Governance and Accountability for Results" (p. 10). The actions the government will take have little to do with student learning or engagement. The report outlines a plan to amalgamate 37 elected regional school boards into 15 regions, dissolve all school boards except for the Division Scolaire Franco-Manitobaine and replace elected trustees and the local governance structure with an unelected Provincial Education Authority that will report directly to the Education Minister. In the name of loving other people's children, Manitobans are being asked to swallow these profoundly antidemocratic changes.

Pillar three of the reform effort is labelled *Future Ready Students*. This section of the document states, "Pathways to further education and employment must reflect student and employer needs" (p. 17). Students are reframed as future employees and the ontological supremacy of human life, and the measurement of achievement is reported through the mouth of Homo Economicus. It is *slippery talk*. Right-wing politicians frame the centralization of power under the guise of meeting the needs of individual students, increasing accountability and measuring student success to ensure taxpayers are getting the best bang for their buck. To argue against the draconian changes, you have to avoid a discursive trap which entices you to argue from within their neoliberal framework. *Don't you want students to be successful? How can you be against modernizing the system when you yourself have said the system doesn't work in the best interest of all families?*

The entrepreneurial culture in education gives primacy to continuous improvement, increasing workloads and the datafication of curriculum and pedagogy. When we become critically conscious of problematic entrepreneurial narratives, we become more adept at resisting the numbing effects and the selective inattention the force of neoliberalism exerts. Ecolinguistics, then, is integral to EELR, because it exposes the assumptions, behaviours and fantasies which are driving the ecological challenges we face. Each time you write an ecosophical counternarrative through EELR, it changes the nature of your involvement in the relational field, creating the possibility for Others to do the same. As our counternarratives coalesce, it will reconfigure the taken-for-granted realities of living outside the Earth's carrying capacity, which makes EELR at its core an ameliorative act of collective resistance.

References

Abraham, N., & Torok, M. (1972). Mourning or melancholia: Introjection versus incorporation. In Abraham, N., & Torok, M. (1994). *The shell and the kernel: Renewals of psychoanalysis.* (pp. 125–137). (Vol. 1). (N. T. Rand, Trans.). University of Chicago.

Benjamin, J. (2018). *Beyond doer and done to: Recognition theory, intersubjectivity and the third.* Routledge.

Blackmore, J. (2006). Social justice and the study and practice of leadership in education: A feminist history. *Journal of Educational Administration and History*, 38(2), 185–200. https://doi.org/10.1080/00220620600554876

Blackmore, J. (2013). A feminist critical perspective on educational leadership. *International Journal of Leadership in Education*, 16(2), 139–154. https://doi.org/10.1080/13603124.2012.754057

Boin, A., 't Hart, P., Stern, E., & Sundelius, B. (2017). *The politics of crisis management: Public leadership under pressure.* (2nd ed.). Cambridge University Press.

Bromberg, P. (2011). *The shadow of the tsunami and the growth of the relational mind.* Routledge.

Farrell, A. J. (2020). *Exploring the affective dimensions of educational leadership: Psychoanalytic and arts-based methods.* Routledge.

Haraway, D. J. (2016). *Staying with the trouble: Making kin in the chthulucene.* Duke University Press.

IPCC (2018a). Summary for policymakers. In: *Global warming of 1.5°C. An IPCC special report on the impacts of global warming of 1.5°C above pre-industrial levels and related global greenhouse gas emission pathways, in the context of strengthening the global response to the threat of climate change, sustainable development, and efforts to eradicate poverty* [Masson-Delmotte, V., P. Zhai, H.-O. Pörtner, D. Roberts, J. Skea, P.R. Shukla, A. Pirani, W. Moufouma-Okia, C. Péan, R. Pidcock, S. Connors, J.B.R. Matthews, Y. Chen, X. Zhou, M.I. Gomis, E. Lonnoy, T. Maycock, M. Tignor, and T. Waterfield (eds.)]. *World Meteorological Organization, Geneva, Switzerland*, 32 pp.

IPCC (2018b). Global warming of 1.5°C: headline statements from the summary for policy makers. https://www.ipcc.ch/site/assets/uploads/sites/2/2019/06/SR15_Headline-statements.pdf

Khalil, D., & DeCuir, A. (2018). This is us: Islamic Feminist school leadership. *Journal of Educational Administration and History*, 50(2), 94–112. https://doi.org/10.1080/00220620.2018.1439904

Lipsky, M. (2010). *Street-Level bureaucracy: Dilemmas of the individual in public services*. Russell Sage Foundation.

Lugg, C. A. (2003). Sissies, faggots, lezzies and dykes: Gender, sexual orientation, and a new politics of education? *Education Administration Quarterly*, 39(11), 95–134, https://doi-org.berlioz.brandonu.ca/10.1177/0013161X02239762

Lugg, C. A., (2017). Skipping toward seniority: One queer scholar's romp through the weeds of academe, *International Journal of Qualitative Studies in Education*, 30(1), 74–82. doi: 10.1080/09518398.2016.1242813

Manitoba Government (2021). *Better education starts today: Putting students first*. https://manitoba.ca/asset_library/en/proactive/2020_2021/better-education-starts-today-report.pdf

Marshall, C., & Sogaard Nielsen, A. (2020). *Motivational interviewing for leaders in the helping professions: Facilitating change in organizations*. The Guilford Press.

Marshall, C., & Young, M. (2013). Policy inroads undermining women in education. *International Journal of Leadership in Education*, 16(2), 205–219, doi: 10.1080/13603124.2012.754056

Miller, W. R., & Rollnick, S. (2009). Ten things that motivational interviewing is not. *Behavioural and Cognitive Psychotherapy*, 37, pp. 129–140.

Miller, W. R., & Rollnick, S. (2013). *Motivational interviewing: Helping people change*. (3rd ed.). Guilford Press.

Ogden, T. H. (1994). *Subjects of analysis*. Aronson.

Portelli, A. (1988). Uchronic dreams: Working class memory and possible worlds. *Oral History*, 16(2), 46–56

Rollnick, S., & Miller, W. R. (1995). What is motivational interviewing? *Behavioural and Cognitive Psychotherapy*, 23. pp. 325–334.

Stibbe, A. (2021). *Ecolinguistics: Language ecology and the stories we live by*. (Second ed.). Edition. Routledge.

Stoknes, P. E. (2015). *What we think about when we try not to think about global warming: Toward a new psychology of climate action*. Chelsea Green Publishing.

Tooms, A. K., Lugg, C. A., & Bogotch, I. (2010). Rethinking the politics of *fit* and educational leadership. *Education Administration Quarterly* 46(1), 96–131. doi: 10.1177/1094670509353044

Warren, R., Price, J., Graham, E., Forstenhaeusler, N., & VanDerWal, J. (2018) The projected effect on insects, vertebrates and plants of limiting global warming to 1.5°C rather than 2°C, *Science*, 360(6390), pg. 791–795. doi: 10.1126/science.aar3646

Winton, S., & Pollock, K. (2016). Meanings of success and successful leadership in Ontario, Canada, in neo-liberal times. *Journal of Educational Administration and History*, *48*(1), 19–34. https://doi.org/10.1080/00220620.2015.1040378

Woodbury, Z. (2019, January 31). Climate trauma: Toward a new a new taxonomy of trauma. *Ecopsychology*. https://www.liebertpub.com/doi/full/10.1089/eco.2018.0021

12 Teacher Education on the Edge of the Anthropocene

Teachers are responsible for the safety and well-being of their students. In the field, we use the Latin term *in loco parentis* to refer to an educator's legal duty to protect the children in their care. This important work can look like ensuring there are no peanut butter sandwiches in lunch boxes to prevent a child with peanut allergies from going into anaphylactic shock or comforting a child after being excluded from a game of soccer baseball. We know in our bones, teaching requires open arms and a passionate commitment to protect the dignity and worth of all children. That said, as I become increasingly conscious of the breadth and depth of the climate breakdown, the moral and ethical obligations of what it means to love and care for other people's children has taken on a new meaning. More specifically, the deluge of bad news compels me to ask with greater urgency: Does the climate crisis demand educators extend their call to teach to ensure livable lives for children yet to come? I recognize the question is an old one. Many Indigenous cosmologies are rooted in intergenerational commitments to all Beings on Earth, but due to Canada's colonial history, in loco parentis remains stuck in a legal construct in many teacher education programmes. The colonial frame of a teacher's duty of care reinforces a reductive understanding of the intergenerational commitments inherent in the work of teaching.

In this chapter, I explore what it would look like in teacher education programmes to take the intergenerational obligations of *in loco parentis* to heart in response to the climate crisis. I begin by framing an image of ecosophical teacher education in reference to a walk I took along the Arlington Street Bridge in Winnipeg a number of years ago. Using a couple of key lessons from the walk along the bridge, I make a relational psychoanalytic turn to breathe life into the emotions, dreams, interrupted words, silences and thought fragments infusing curriculum and pedagogy. Then, I draw a more detailed picture of relational teacher education, one in which teacher candidates develop confidence in guiding classroom inquires open to the unbidden. In this new evocation of ecosophical teacher education, I imagine teacher candidates working with compassionate mentors who weave a critical pedagogy of place as sojourners who value and learn from their interdependence among all biological and cultural systems.

A Psychoanalytic Turn on the Arlington Street Bridge

When I pick up my camera, I feel more in tune with my surroundings. The world becomes more vibrant somehow because photography affords me the opportunity to uncover new understandings of the life affirming and problematic relationships among people and other Beings. Approximately six years ago, I grabbed my camera and went for a long walk across the Arlington Street Bridge in Winnipeg, Manitoba. The view of the bridge is captured in Figure 12.1. During the walk, I was taken aback by the thousands of pigeons nestled on top of the train cars in the yard below the bridge. I spent hours watching how the birds interacted with the people, machines and sounds produced by the activity in the yard.

Walking outdoors can foster an affinity for the natural world. Fallen logs become places to study interesting insects, a hurdle to jump over or even a magical home for fairies. Learning becomes more vibrant outside because we are thrust into a multi-sensory pedagogy of place in which the smells of bus fumes or fields of clover imprint themselves in the layers of learning encounters. More importantly, learning outside is an opportunity for children, powered by their own curiosities, to take the lead. As an example, my sons who walked the bridge with me that

Figure 12.1 A View from the Arlington Street Bridge

day directed my attention to the flashes of indigo, pink and silver on the pigeons' wings. They also learned and performed various pigeon movements while they watched the birds jostle for position on top of the train cars. My children became my guides, pulling my attention to a pigeon aesthetic and the lessons the birds had to teach. In this way, the walk along the Arlington Street Bridge was a reminder of the necessity for teachers and educational researchers to amplify pedagogies that build upon the knowledge and visuality of children as they engage in experiential learning adventures.

With my camera, I was able to capture aspects of the relationship between the living and non-living forces in the yard. For example, I took pictures of various birds responding to screeches of metal on metal. Likewise, educators exist in a web of relations with their students. Learning exerts a series of relational vibrations, and an important aspect of teaching is to attune to the effects of those waves on Others. In this way, pedagogy cannot be reduced to the acquisition of new methods and strategies. Instead, pedagogical praxis calls teachers to listen more intently and to respond more intentionally to the emotional resonances infusing educative spaces.

To attune to the more-than-human world, value must be placed on nurturing brave and contemplative encounters, encounters in which young people are encouraged to seek beauty in themselves, Others and in the Land, Water and Skies. It requires teachers and students to be open to the arrival of sensations, questions or ideas beyond their wanting and doing. The cultivation of this type of pedagogical disposition, one accepting of unbidden beauty, necessarily privileges slow and deep learning over technocratic learning structures, individualism and the expedient targeting of learning outcomes. Furthermore, this pedagogical approach is ecosophical in its orientation because it cultivates a sensitizing visuality, one in which an appreciation and concern for other Beings is fundamental to becoming an educated person. Maxine Greene (2001) makes the case beautifully for why it is important walk wideawake in the world and how *wideawakeness* exists in relation to seeing the beauty in other Beings:

> You have to be fully present to it - to focus your attention on it and, again, to allow it to exist apart from your everydayness and your practical concerns. I do not mean that you, as a living person with your own biography, your own history, have to absent yourself. No, you have to be there in your personhood, encountering the work much in the way you encounter other persons. The proper way to encounter another person is to be open to them, to be ready to see new dimensions, new facets of the other, to recognize the possibility of some fresh perception or understanding, so you may know the other better, appreciate that person more variously. This is, actually,

how we ordinarily treat each other as persons. We do not treat each other as case histories, or instances of some psychological or sociological reality – not, that is, in personal encounters. Nor do we come up against each other as if the other were merely an inanimate object, incapable of reciprocation. There are analogues between this and encounters with works of art, especially in the readiness for fresh illumination, in the willingness to see something, to risk something unexpected and new.

(2001, p. 53)

Just before I packed my camera to head home from the Arlington Street Bridge, one of the pigeons unexpectedly popped up in front of my lens. The moment is captured in Figure 12.2. I was startled by the bird's sudden appearance and I almost lost hold of the camera. Immediately after the bird flew away, I thought about learning as a series of sudden intrusions. There have been countless moments when something I read took me by surprise because an idea unsettled some of my strongly held beliefs about teaching. It is one of the most troubling and liberating aspects of learning, assumptions in educative spaces can "pile up like casualties and treasures, both" (Alvarez, 2011, p. 94). Often these emotional collisions happen in spite of what a learner is trying to learn or what the teacher is trying teach.

Uncertainty in a pedagogical encounter is often interpreted as a challenge to authority, but like Greene suggests, to risk something unexpected

Figure 12.2 A Pigeon Soars

is at the heart of education. This is both foundational and unnerving, as teachers and students are always affected by the unknowable other and the object of study (Britzman, 2003; Britzman, 2006; Britzman, 2009). Consequently, I have come to understand education as affectively charged and the attunement to telluric emotions *as education*. Pedagogy, then, is becoming more attentive to students as they encounter new ideas and the art of becoming increasingly focused on what escapes students' awareness but still manages to influence their actions. Teachers are always at work in the realm of disparities, awkward silences and fleeting impressions while articulating partial meanings of felt experiences. And in pursuit of mutual recognition, teachers deliberately open space for children to vocalize curiosities, particularly in the moments when the teachers' desire is to close the space (Britzman, 2013) in service of their own egos or needs for control.

Freud called psychoanalysis and education impossible professions, forever doomed to failure. In both fields, the enigmatic unconscious has its way and the significant relationships (analyst/analysand and teacher/student) are rife with confusion about dependency. In both fields of study, fantasies of what could be are quickly translated into echoes of what has already happened. As an illustration, it is difficult to escape being transported back to the moments when we awkwardly grasped the fat red pencil with our tiny fingers to print our names under the watchful eye of our teacher whenever we try and think pedagogy anew. In this sense, educators and educational researchers are always stepping into the third act of a play in which all the characters are reciting lines from different productions. Consequently, to play within the paradox of articulating a new ecosophical orientation in teacher education in a field that Britzman (2013), in reference to Bion's work (1993), called "...a thought without a thinker", I was compelled to paint a picture of teacher education *as if* ecosophical passions have already ruptured the field.

An Ecosophical Teacher Education Programme

I took inspiration from the Intercept's short film *A Message from the Future with Alexandria Ocasio-Cortez*. The film is a collaboration among author and activist Naomi Klein, congresswoman and scriptwriter Alexandria Ocasio-Cortez, illustrator Molly Crabapple, co-directors Kim Boekbinder and Jim Batt and writer and documentary film-maker Avi Lewis. Set a few decades into the future, the seven-minute film narrated by Ocasio-Cortez invites us to imagine the world as if we radically changed course, faced the climate crisis and then managed to strike a Green New Deal (Klein, 2019). I, too, would like to invite you to imagine a future in which teacher education in the Anthropocene assumes that a dystopian future is not a given. Grating against the images and text of *A Prairie Elegy for the Discerning Consumer* in Chapter 3, the next section tilts away from the post-political machinations of the elites and away from the structural critiques of progressives that keep us locked and yelling from within oppressive frameworks and barriers to change. I offer an image from the future *as if* an ecozoic consciousness (O'Sullivan, 1999) has already ruptured teacher education programmes. I group the tenets ecosophical teacher education under ten intersecting pedagogical praxes.

A Pedagogy of Compassionate Witnessing

Due to the psychological impacts of climatological precarity, in the early 2030s the number of children and youth dealing with anxiety and depression spiked. Teacher education programmes began to take the emotional dimensions of learning seriously and it became a moral imperative to understand oneself as a *compassionate witness* in the classroom. Students now study Winnicott's work to better understand classrooms as holding environments. They work alongside climate social workers to think deeply about what it means to bear witness to the grief, anger and anxiety children express about the uncertainty of their future. They learn the attunement to a child's feelings after a traumatic climate event makes it less likely for the traumatic experience to produce long-term effects. Furthermore, teacher candidates see first-hand how the social ties among students and the wider community are strengthened when pedagogy is directed to the education of a relational self. As a result, teacher candidates spend a great deal of time in the field and on campus reflexively learning how to be more capable of caring for diverse Others in precarious times.

The turn to the Other has become an organizing principle of the method and matter of teaching. A focus on their own compassionate witnessing helps teacher candidates create experiences in which children learn to take responsibility for the Other, even when the Other cannot or will not return the gesture. Vulnerability and compassionate witnessing are built into pedagogical structures. Teacher candidates are introduced to several drama activities to foster the verbal and non-verbal language of attunement in their classrooms. One of the common exercises they learn to facilitate in the first year of their programme is called *Hypnotism* (Boal, 2002; Boal, 2006). In the game, two people stand facing one another. Partner A raises her hand. Partner B focuses all of his attention on Partner A's hand. Trying to keep the same distance between his nose and Partner A's hand at all times, Partner A moves them about the room. A few minutes into the game, the facilitator asks the players to switch roles. After the game ends, the facilitator asks the players to report on the tensions that emerged in the game in relation to leadership, followership, trust and mutual recognition.

Compassionate witnessing extends to the more-than-human world. Teacher candidates critically examine the anthropocentric assumptions they inherit, particularly in regards to Western notions of progress and the benefits of technology. In the course *A Historical Analysis of the Complex Systems and Structures in Education*, students study the intersections among the COVID-19 pandemic, the climate emergency, colonialism and the challenges to democracy between the tumultuous years of 2015–2030. In another course called *Education and Trajectories of Freedom*, attention is given to the dynamic fields of existence, assemblages and trajectories (Greenhalgh-Spencer, 2014) rooted in and emerging from the local ecological schooling landscape.

A Pedagogy of Deep Ecology

A strong emphasis is placed on understanding curricula in the emergence of an appreciation and love for the uniqueness of Others and the Land, Water and Skies. As a place to begin, teacher candidates work closely with elders and knowledge-keepers in their communities to understand their ecosophical orientations to pedagogy and Indigenous cosmologies. Local members of Decolonizing Water Initiative work with teacher candidates so they can enhance their capacity for two-eyed seeing (Decolonizing Water Project, n.d.). Additionally, students critically reflect on eight of Naess's (2008) deep ecology principles:

1. All living beings have intrinsic value.
2. The richness and diversity of life has intrinsic value.
3. Except to satisfy vital needs, humans to not have the right to reduce the diversity and richness.
4. It would be better for humans if there were fewer of them, and much better for other living creatures.
5. Today, the extent and nature of human interference in the various ecosystems is not sustainable, and the lack of sustainability is rising.
6. Decisive improvement requires considerable changes: social-economic, technological and ideological.
7. An ideological change would essentially entail seeking a better quality of life rather than a raised standard of living.
8. Those who accept the aforementioned points are responsible for trying to contribute directly or indirectly to the necessary changes (p. 28).

Using Naess's work in the context of teacher education, the candidates make thoughtful judgements about student success in relation to inner peace, fostering a deep connection with the natural world and collective well-being.

A Pedagogy of Feeling

Teacher educators engage in self-analysis to explore how their own messy feelings get in the way of the unconditional acceptance of children. Recently, during a practicum placement in a grade two classroom, a student-teacher told his advisor that he noticed his patience with the children ran short each time the advisor was in the room conducting a formal observation. The teacher candidate's worries about being judged for not "properly managing a noisy class in the eyes of his advisor" sucked the spontaneity and joy out of his engagement with the students, leaving the desire to control and manage their behavior in his wake. Throughout their programme, teacher candidates engage in these types of reflective conversations to become more practiced at observing their own feelings and the feelings of Others with curiosity. They learn about the relational psychoanalytic concepts of mutual recognition and enactments and discuss how difficult it is to cultivate and sustain moments of mutual recognition. Advisors encourage student teachers to share how they feel when the behavior of those in their care makes their cheeks flush or their blood boil. Using historical documents in digital archives, documentaries, literature, film and plays to paint the emotionally complex history of schooling, teacher candidates interrogate instances of negative emotions fuelling harmful actions done in the name of caring for other people's children.

Teacher candidates are acutely aware of the affective dimensions of the climate crisis. Many students they work with during their field experiences express anger at older generations for not acting with greater urgency and determination to mitigate the damage. The sentiment "it didn't have to be this way" is more acutely felt and vociferously expressed, when schools are closed during severe weather events. In response, teacher education programmes develop and share resources to support students, teachers, school leaders and community members to mitigate the psychological pain in the aftermath of a serious climate event. As an illustration, one teacher candidate made and uploaded a video on the Faculty of Education's resource hub to share an experience she had working in a school after an F2 tornado severely damaged parts of the community. In the video, the teacher candidate describes how the school organized a World Café (theworldcafe.com) to open a dialogical space for students and families to mourn the damage to the community and to make sense of the increasing intensity and frequency of spring storms. The World Café began with introductions followed by an acknowledgement of the damage the tornado had caused to the school community. The facilitator posed a question to invite discussion in small groups about their shared experience, and after 20 minutes, each small group shared its insights with the large group. Visual representations of each group's discussion were posted around the room. In the video, the teacher candidate explained how she collated the information under the themes of *impacts, resilience factors, actions* and *community building* so the learning from the event could be shared with other community members who were not in attendance. Her video is a powerful example of a school community processing telluric emotions in the aftermath of a severe weather event.

To cultivate acts of self-compassion, a concerted effort is placed on acknowledging one's feelings and the feelings of Others without judging them. Teacher candidates are taught how to recognize the physical manifestations of climate anxiety in the flushed cheeks, teary eyes or gritted teeth of their students. As a corollary, instructors encourage teacher candidates to think critically about how and why different affects register publicly in classrooms and how publicly registered affects determine who is allowed to speak and under what conditions about climate change. In addition, teacher educators invite teacher candidates to experiment with pedagogical practices that open space for embodied eco-grief and *eco-jouissance* to be processed in classrooms. As an extension of living curriculum each year, the teacher candidates organize a cellphilm festival (MacEntee et al., 2016) to express reverence for the Land, Sky and Water sustaining their lives.

A Pedagogy of MultiVocality

University administrators followed through on their commitments to recruit, support and retain linguistically and culturally diverse cohorts of teacher candidates to ensure the teaching profession is representative of the children and families in all school communities. Teachers as diverse and polysemic public intellectuals, amplify the multivocal potential of Others so young people expand their notions of diversity and come to value the rich tapestry of cultural and biodiversity in their school communities. Monocultures in classrooms and in the wider environment inhibit the flourishing of life. In this vein, teacher candidates learn to cultivate a polyphonic (many-voiced) attunement to the Land, Water, Animals, Plants and Others to centre what the more-than-human world has to teach us about climate change and living within the Earth's ecological limits.

The multivocality of young people's lives is regarded as a sacred entry point when responding to the multifaceted challenges global warming presents. It is challenging to engage in complex and emotionally charged eco-justice activism. To do this difficult work, teachers foster a dialogical self, one who engages Others relationally to co-construct a shared ecosophical identity. Sometimes this requires a teacher to unsettle the hierarchical production of knowledge within the classroom so children and youths can authentically shape what and how they learn. By learning to co-design inquiry projects with students during their field experiences, teacher candidates practice coaching young people to identify, hone and use their own unique gifts to make a positive difference in the lives of Others. At the end of each school season, teacher-candidates and their students share what they have learned through their inquiry projects and how their learning will address some aspect of one of the most pressing issues facing their more-than-human relatives.

Cultural practices claim space in our imaginations and form the foundation of our individual and collective passions and inquiries. That said, teacher candidates claim a moral and ethical obligation to continually expand the space for the stories of children, youths and families who have been historically marginalized in school communities (Rodriguez, 2020). During their *eco-aesthetics* course, student-teachers critically reflect on how monocultural pedagogies limit the individual and collective potential of children, and they also study a diverse set of works from poets, playwrights, filmmakers and narrative photographers to expand the poetical capacities of their creative self. Later in the course, they inquire into what it means for teachers to become storytellers and culture-makers in the age of the Anthropocene. If teachers expect students to be *creatives*, they too need to practice, share and then weave their own aesthetic contributions into the cultural tapestry of the school community.

A Pedagogy of Dialogic Fascination

Teacher candidates recognize conflict surfaces whenever students engage in conversation about contested issues. Instead of responding within a disciplinary mindset to expressions of messy emotions, they learn to seek out the underlying feelings precipitating the animus rather than choosing to sit in judgement. Becoming-teachers draw upon the many narrative techniques they experience in their courses to critically reflect on the communication patterns they inherited from the formative relationships in their own lives. To further support them in becoming more artful communicators, course instructors model *dialogic fascination* in their own classes. Dialogic fascination describes a state of wonder and curiosity as one receives and interprets the Other's response within an encounter. In the context of the classroom, dialogic fascination calls teachers to deeply listen, embody a spirit of generosity and remain open to the unexpected as students interpret the world through the stories that have shaped their lives.

One of the drama exercises instructors use to help teacher candidates strengthen their skills in this area is called *The Teacher Thought Bubble*. To begin the exercise, an instructor asks a volunteer to describe a moment when they were engaged in an emotionally charged conversation with a student or a colleague during their most recent field experience. The volunteer storyteller is informed they cannot name the other parties involved nor are they allowed to disclose any details which could identify others. Then, the storyteller sits on a chair facing the rest of their classmates and describes a specific moment when she felt uncomfortable during the conversation in question. One at a time, students volunteer to stand behind the storyteller and share an emotional reaction to what they heard, as if they were a voice in the storyteller's mind. The technique not only opens space to make collective sense of complex emotions, it becomes a live demonstration of syncing (in content and method) the co-creation of a shared reality in which all people within the learning encounter feel recognized and respected.

Teacher candidates are introduced to other communication exercises to help them, and eventually their future students, divest from problematic repetitions of relational patterns that impact one's ability to recognize the Other. They learn not to ignore communication difficulties. When people habituate to unhealthy communication patterns, it negatively impacts future encounters in which conflict arises. As an example, teacher candidates are asked to practice acknowledging and tolerating their own feelings of nervousness during the silences that follow a question. Some people need to percolate longer than others, so it is unhelpful to judge people negatively for language that remains on the way.

A Pedagogy of Self-Analysis

At the beginning of their programme, teacher candidates consider how their formative relationships shape their relationships to Others to gain a more fulsome understanding of their own moral and political attachments. During their *Teacher Identity* class, one teacher candidate is finally able to connect her struggles with seasonal depression to the absence of frozen rivers and lakes. Throughout her childhood, she was drawn to moving water as a method to quell her worries and ground herself in the present moment. Rising temperatures have elongated the time of running water in her home community. She is now making sense of the ambivalence and guilt she feels when she finds herself relaxing on the banks of a river that should already be frozen. By the time she enters the programme capstone course, she judges her ambivalence less harshly and uses the realization to be less judgemental in the presence of Others who ignore the terrible reality underneath the weather reporter's toothy grin and sunny forecast.

A Pedagogy of Mitigation

Throughout their programme, teacher candidates practice designing learning experiences focused on local and global multi-species impacts related to air quality, food safety, severe weather events or rises in sea levels. They become more confident, composing inquiry questions in age-appropriate terms while maintaining a focus on building student agency. As an example, a teacher candidate, in her last practicum placement, studied the health indicators of forest ecosystems with her students. The class visited a nearby forest several times throughout the term. One group of students studied tree growth indicators. They looked at the balance of new growth in relation to tree mortality. A second group of students focused on lichen abundance. This group taught their classmates that lichens are extremely sensitive to environmental stressors and changes to forest structures. A third group of students took soil samples to determine how acidic or alkaline the soil was, while a fourth group studied tree species diversity. After the students finished collecting their data, they presented their findings to the other multi-age classes, the parent council and to the city council.

The students' research projects led to the development of the Friends of the Forest partnership. The partnership consisted of the class, five forestry management students studying at the local college, one member of the city council and a small group of seniors who cared deeply about the changes they observed in the health of the forest. The Friends of the Forest learned how to amplify their knowledge about the gradual decline of the forest's health in ways that increased the public's attention to smaller decision-making trigger points. If raging wildfires are the guttural screams of a forest, the forest in their own community was beginning to shout for help. Responding to the call, this group of eco-conscious citizens developed several teach-in strategies to communicate with other community members about why it was important to become stewards of the forest. As a result of the activities of the Friends of the Forest, critical interventions and rewilding efforts were made before there were more disturbances to the forest's well-being.

A Pedagogy of Social Dreaming

Climate change is an alchemy of affects and social attachments. The pursuit of an ecosophical orientation to living amid the Earth's decreasing habitability requires one to be able to share the moral, political, somatic and aesthetic poetics of their dreams. Using guided imagery and visualization exercises, student-teachers rehearse using generative methods to analyze and solve novel problems with students. One teacher candidate had an opportunity to work closely with the school psychologist who facilitated Climate Cafés with students and members of the wider community. During his practicum with the school psychologist, she taught him about Alfred Adler's (2014) *as if technique*. During the first term of his practicum placement, he used the technique when he noticed some of his students struggling to see themselves as powerful people who could make a difference in the world. After one of his grade five students said,

"We're just kids. Nobody listens to us anyway", the teacher candidate responded, "You might act as if it's a given that adults don't listen to kids. I am curious to know how you think your life would be different if adults *really* listened to your concerns. What would you sound like in a conversation with an adult if they did take you seriously? What would the adult sound like? How would your relationships change if the adults in your lives took you seriously? What might happen if you started acting *as if* they did take you more seriously?"

The COVID-19 pandemic and the climate emergency amplified the need for schools and other institutions to change in unprecedented ways to escape the anthropocentric delusions responsible for the pain and suffering of billions of Beings on the planet. Extreme weather events, food shortages, wildfires, mass climate migrations and breakdowns in transportation systems were some of the shocking "lessons" the two intertwining crises wrought in communities all over the world. Awe-inducing and often psychically numbing (Lifton, 2017), these events put the effectiveness of hyper-techno fetishizing in standardized curricula into question. One of the methods educational researchers use to unhook participants from their fantasies of control is a process called *Social Dreaming* (Manley, 2020). Working in groups of 10–25 people, participants share their dreams and have them placed in a dreaming matrix. The process takes a crisis like a pandemic or the climate emergency and distills it into condensed images to make what is unimaginable a part of a group's visuality. Metaphors, poetry, radical thoughts and the emotional dimensions of social learning emerge organically as dream threads from each individual weave a shared tapestry. The hyper-object (Morton, 2013) is distilled by constructing new meaning from the dream parts of individual actors. Unburdened from the logics of cause and effect, the participants' free associations circulate among the participants in the room producing innovative and evocative aesthetic assemblages of the weirding climate.

A Pedagogy of Land-Based Learning

Teacher educators, in collaboration with teacher candidates, Indigenous knowledge keepers, partner teachers, student activists, school and division leaders and other community experts (zoologists, entomologists, ornithologists, botanists etc.), collaborate to provide transdisciplinary field experiences to prepare teacher candidates to facilitate land-based inquiry projects. Land-based learning is foundational to ecosophical pedagogy because:

> We may be said to be in, and of, nature from the very beginning of ourselves. Society and human relationships are important but our own self is much richer in its constitutive relationships. These

relationships are not only those we have with other humans and the human community (I have elsewhere introduced the term *mixed community* to mean those communities in which we consciously and deliberatively live close together with certain animals), but also those we have with other living beings".

(Naess, 2008, p. 82)

Consequently, the *Wilding-Field-Experience* is a fundamental programme requirement. During this multi-year transdisciplinary field-led course, teacher candidates become curriculum-makers and capable guides in wild places. In school communities with limited access to wild places, teacher candidates, students, field supervisors and community partners design and implement projects to rewild tiny areas of land in close proximity to the school. Several wildlife rejuvenation projects are completed each year as a result of the course. The work of teacher candidates has led to the rewilding of land around playgrounds, the construction of outdoor classrooms and increases in the practice of urban gardening.

By situating student and teacher learning within a critical pedagogy of place, researchers in the field of education have noted several benefits for children. In studies of outdoor play-based learning environments, young people are more capable of solving complex problems, they are more compassionate when trying to resolve conflicts and they become significantly more knowledgeable about the biological diversity in their school communities. In relation to their own lived experiences, teacher candidates enhance their own *embodied knowing* of the natural world. It is delight in the shaping shifting of clouds, getting tickled by a head of wheat as it grazes your fingertips, being awestruck by the vastness of the ocean or feeling a rush of life as you breathe in the smell of the forest floor after a long rain.

In the first half of the *Teacher Identity* course, students study the work of Janet Cardiff and her sound sculpture called Forty-Part Motet. In the piece, she reimagines *Spem in Alium* by the 16th-century English composer Thomas Tallis (National Gallery of Canada, 2013). Cardiff recorded 40 individual choir voices and plays each voice through one of the 40 speakers placed intentionally around the gallery. Using Cardiff's haunting Forty-Part Motet as inspiration, teacher candidates create their own soundscape from a place they spent some of their formative years. Collectively, the work makes a cabinet of curious sounds, containing the embodiment of the soundtracks to their teaching lives. They wrap their soundscapes of learning around the rigorous study of relational approaches to systems thinking in the natural world, critically reflecting on the dangers of anthropocentrism, and learning how children's emotional well-being is rooted in the health and stability of the ecosystems.

A Pedagogy of Interdependence

There once was a myriad of hierarchical relationships strangling relationality in the field of education, particularly in the area of formal assessment. One of the triggers of the assessment revolution were a series of studies published about the negative impacts grading had on students. Grades squelched curiosity about topics, decreased willingness to take intellectual risks and impeded deeper thinking when students were more focused on external reward mechanisms. Thankfully, there were several substantial changes to evaluation practices across school systems. As relational pedagogical practices took root in faculties of education approximately 20 years ago, there was a concerted effort to move from grading to ungrading across the K-12 public school system. Proponents maintained that if educators were to teach relationally, meaning they would privilege mutuality, adopt a stance of curiosity before passing judgement, take the affective dimensions of learning into serious consideration and move beyond viewing a child as *omnipotently controllable* (Stern, 2015), then letter grades and other unreliable schemes of institutional control needed to be eliminated (Blum, 2017; Freire, 2018; Schinske & Tanner, 2014).

When letter grades and traditional report cards were first eliminated, fierce criticism was leveled at faculties of education, teachers and school administrators. Many parents and caregivers wanted to know how they would be able to rank their child's performance against the achievements of other children. Five years into the assessment revolution, advocates and progressive educators wondered if they would have to retreat. Thankfully, teachers were able to amplify the benefits of ungrading in relation to student learning and engagement. They shaped their own research which revealed ungrading promoted self-evaluation and metacognitive thinking. It encouraged students to reflect more frequently on their own progress and it treated students as experts in their own learning. Climbing out from under the oppressive weight of standardized exams, prescribed learning outcomes and linear notions of curriculum, teachers had more time for rich and rigorous conversations with individual and groups of students about how their education was helping them to become the finest expressions of themselves in the face of the Other.

To model interdependence in the context of education, teacher candidates observe most of their instructors team-teach. The lore of the lone wolf professor and the hyper-competitive culture of the academy are viewed as cautionary tales from a time when external grants and peer-reviewed publications were more valued than fostering meaningful relationships within sustained inquiries. Selfies of exhausted professors who smelled of airport fumes at academic conferences withered away on academic Twitter. Now, teacher educators are rooted in their own community contexts. Many engage their student-teachers in community-based research projects to study climate education, land-based learning, stewardship and eco-restoration as moral imperatives in spaces of curriculum and pedagogy.

References

Adler, A. (2014). *The practice and theory of individual psychology*. Lushena Books.
Alvarez, J. (2011). *That moment. poem in the woman i kept to myself*. A Shannon Ravenel Book.
Blum, S. D. (2017). The Significant Learning Benefits of getting rid of grades. *Inside Higher Ed*. Retrieved from https://www.insidehighered.com/advice/2017/11/14/significant-learning-benefits-getting-rid-grades-essay
Boal, A. (2002). *Games for actors and non-actors*. (A. Jackson, Trans.). Routledge. (Original work published in 1992).
Boal, A. (2006). *The aesthetics of the oppressed*. Routledge. Retrieved April 14, 2010, from http://lib.myilibrary.com.proxy1.lib.umanitoba.ca/browse/
Britzman, D. (2003). *Practice makes practice: A critical study of learning to teach*. State University of New York Press.
Britzman, D. P. (2006). *Novel education: Psychoanalytic studies of learning and not learning*. Peter Lang.
Britzman, D. (2009). *The very thought of education: Psychoanalysis and the impossible professions*. State University of New York Press.
Britzman, D. P. (2013). Between psychoanalysis and pedagogy: Scenes of rapprochement and alienation. *Curriculum Inquiry*, 43(1), 95–117. https://doi.org/10.1111/curi.12007
Decolonizing Water Project (n.d.) Water is language: Kegedonce John Burrows shares some teaching about Nibi (water) from his home community of Neyaashiinigmiing. https://www.waterteachings.com/water-is-language
Freire, P. (2018). *Pedagogy of the oppressed*. (50th Anniversary ed.). Edition). Bloomsbury Academic.
Greene, M. (2001). *Variations on a blue guitar: The Lincoln center institute lectures on aesthetic education*. Teachers College Press.
Greenhalgh-Spencer, H. (2014). Guattari's ecosophy and implications for pedagogy. *Journal of Philosophy of Education*, 48(2), 323–338. https://doi.org/10.1111/1467-9752.12060
Klein, N. (2019, April 17). A Message from the future with Alexandria Ocasio-Cortez. The Intercept. https://theintercept.com/2019/04/17/green-new-deal-short-film-alexandria-ocasio-cortez/
Lifton, R. J. (2017). *The climate swerve reflections on mind, Hope and survival*. The New Press.
MacEntee, K., Burkholder, C., & Schwab-Cartas, J. (2016). *What's a cellphilm: Integrating Mobile phone technology into participatory visual research and activism*. Sense Publishers.
Manley, J. (2020). The jewel in the corona: Crisis, the creativity of social dreaming, and climate change. *Journal of Social Work Practice*, 34(4), 429–443. doi: 10.1080/02650533.2020.1795635

Morton, T. (2013). *Hyperobjects: philosophy and ecology after the end of the world*. University of Minnesota Press.

Naess, A. (2008). *The ecology of wisdom: Writings by arne naess*. A. Drengson & B. Devall (Eds.). Counterpoint.

National Gallery of Canada (2013, January 23). Critically acclaimed sound sculpture opens new NGC@WAG program. https://www.gallery.ca/for-professionals/media/press-releases/critically-acclaimed-sound-sculpture-opens-new-ngcwag-program

O'Sullivan, E. (1999). *Transformative learning: Educational vision for the 21st century*. University of Toronto Press/Zed Books.

Rodriguez, F. (2020). Harnessing cultural power. In A. E. Johnson & K. K. Wilkinson (Eds.), *All we can save: Truth, courage, and solutions for the climate crisis*. (pp. 121–127). One World.

Schinske, J., & Tanner, K. (2014). Teaching more by grading less (or differently). *CBE—Life Sciences Education*, *13*(2), 159–166. https://www.lifescied.org/doi/10.1187/cbe.cbe-14-03-0054

Stern, D. B. (2015). *Relational freedom: Emergent properties of the interpersonal field*. Routledge.

Get Busy Living

I found myself watching more movies when isolating during the COVID-19 pandemic. One of the films I watched more than once was, *The Shawshank Redemption* (Darabont, 1994). The film is about a banker named Andy Dufresne who is sentenced to spend the rest of his life in the Shawshank State Penitentiary for the murder of his wife and her lover. Despite his claims of innocence, Andy spends 20 years in prison. In that time, Andy is befriended by a fellow inmate called Red. In the latter part of the film, after Andy spent two months in solitary confinement, he tells Red he dreams of living in Zihuatanejo, a Mexican coastal town. Red projects his own anxiety onto Andy's expression of hope and exclaims, "Goddamn it, Andy, stop! Don't do that to yourself! Talking shitty pipedreams! Mexico's down there, and you're in here, and that's the way it is!" Andy responds, "You're right. It's down there, and I'm in here. I guess it comes down to a simple choice, really. Get busy living or get busy dying".

As I watched the movie for the second time, I thought about this iconic scene and the line, "Get busy living, or get busy dying" in the context of the climate crisis. I thought about how the seductive forces of late stage capitalism have imprisoned our thinking and poisoned our politics. It appears impossible for some politicians to reference living outside of the *I think therefore I consume* economic imperative. I thought about how expressions of social solidarity, demands for equity and racial justice and calls for a Green New Deal, are framed as "shitty pipedreams" because our thinking is colonized and we are collectively terrified to face our climate trauma. I think about how expressions of optimism push against our subconscious resignation (it is already too late) projected in the apocalyptic films, literature and murder shows we devour. Maybe Andy's retort, "Get busy living or get busy dying" is a distillation of the primal tension living in the Anthropocene presents. What does it mean for educators and educational researchers to *get busy living*? Does an attunement to eco-anxiety and eco-grief hold emancipatory potential or is the recent attention the media is paying to eco-anxiety a luxury of the white and wealthy in the Global North? "Put another way, is climate anxiety just code for white people wishing to hold onto their way of life or get 'back to normal', to the comforts of their privilege?" (Ray Jacquette, 2021, para 2) and if it is a privilege smokescreen for some, can education cultivate greater responsivity to the suffering stranger? Are we ready to confront the ways education is structured so we subconsciously *get busy dying*? What prisons, material and dialogical, need to be torn down if we were to learn to *live relationally*?

Although all learning encounters are affectively charged, biodiversity losses, extreme weather events, rising sea levels and mass migrations will undoubtedly present additional emotional burdens, making educative encounters more complex in years to come. Consequently, educational research must be in part, predicated on its ability to engage with *telluric emotions* because we cannot be healed or studied apart from our emotional connections to the planet. Throughout the text, I tried to illustrate how difficult knowledge about global warming elicits social defence mechanisms such as denial, disavowal and projection, thereby suppressing the recognition and processing of telluric emotions. I went on to argue, that to care for our own and the defensive responses of others, what is needed is a reorientation in the field of education toward ecosophical ends and robust research focused on reimagining classrooms as relational homes for climate dread.

To ameliorate the psychological impacts of climate change, pedagogical praxis must become more attentive to students' emotions as they live and learn in uncertain times. Capitalizing on the power of relational psychoanalysis to address the affective dimensions of the Earth's decreasing habitability, I proposed a significant reorientation in education toward ecosophical ends. Ecosophical education (EE) is grounded in an ethos of mutual recognition, dialogical fascination with the Other, non-violence, democratic participation and compassionate witnessing. That said, educational researchers have an important role to play in studying and disseminating knowledge about attuning to our more-than-human relatives, building a relational home for students, amplifying learning processes that privilege interdependence over individualism, increasing understanding about the emotional dimensions of the climate crisis and fostering resiliency processes.

An intersubjective approach within the field of educational research is urgently needed because the widespread evasion of the climate crisis is fostering epistemic mistrust among young people. Using the work of Abraham and Torok (1971; 1972), I explained how trauma remains hidden from a sufferer's conscious awareness through the process of incorporation. The disavowal of the seriousness of the ecological emergency in the field of education produces traumatizing effects in individuals and across the relational field, and when environmental losses are introjected, it precludes one's ability to mourn. Greenwashing and placations will continue to shake young people's sense of themselves, causing more to lose faith in the safety of their environment. Educational researchers could contribute positively within the emotional registers in education, by centring the voices and concerns of those who are most impacted by the effects of living outside the planet's carrying capacity. Moreover, insights from EER could be used by educators to animate collectivistic approaches to complex transdisciplinary questions.

I extend an invitation to other educational researchers to address the sources of ecological degradation that linger below the surface of the technical and rational discourses shaping research, policies and practices. Together, we can draw urgent attention to the destruction of long-evolving interdependencies and nurture our research affinities and curiosities to become response-able for bending curriculum and pedagogy towards compassion and mutual recognition and away from competitiveness, indignity and suffering. An eco-conscious, decolonized embodiment of *in loco parentis* imparts an ethical obligation for researchers, teachers, leaders and policy makers to ensure liveable parts of the planet exist for future generations. Education can be a force and source of inspiration to enact new ways of living together that are more just, equitable and compassionate. We can begin by inviting our colleagues, and the teacher candidates with whom we work, to look at education anew as we teeter the edge of the Anthropocene. To be sure, the work will tax our individual and collective resiliency processes but it will also strengthen our interconnectedness and nurture a shared purpose to protect the Earth, our home.

Educators, like photographers and psychoanalysts, move with and against the preservation of affects, with the hope of working through an idea, feeling or an echo of identity in the present. Using a feeling-photography to draw attention to climate precarity, biodiversity losses and other planetary boundaries exceeded by human activity, I sought to interrupt the pattern of ecological erasure in the field of educational research. Moving against the dominant selfie-culture, I turned my lens toward the Other to open a research trajectory amid the monstrous aesthetics of the Anthropocene. My theoretical wager is that educational researchers will find research-creation valuable terrain upon which to engage in disobedient thinking about human beings' exertion of power to the detriment of Others. I argued, monstrous feelings embedded in creative acts, pull attention to humanity's role in the unmaking of ecosystems, that is, *the process of becoming monster*, so we may better understand the legacies of our species' biomorphic powers. Research-creation facilitates emotional composting, and through arts-based research, it may be possible to trouble internalized myths about saviour technologies and human dominance in education, while enhancing our ability to bear witness to our own, and the monstrous feelings of Others.

Feeling-photography was a way to reflect on and account for the emotionality embedded in inherited frames of curriculum and pedagogy. I continue to use the energy from creative processes to wrestle with my own ambivalence in the face of difficult knowledge. I have no doubt monstrous feelings will continue to surface in the future when I hear politicians greenwash educational policy or proffer flimsy excuses for climate inaction. On these days, I will try and be more aware of what is abundant instead of what is missing. Thankfully, through research-creation, I am continually reconnected with my more-than-human-relatives. A picture of a bear cub eating a raspberry, a photo of a ladybug on a child's fingertip, an image of a graceful tern soaring above the lake, they are all awe-inspiring reminders of the beauty that surrounds us.

Unlocking emotional crypts in the affective field of relations, requires an assertion of the value of one's life and lives of Others. I queried the mechanisms the privileged (gender, race, class, sexual orientation) use, to collude in the maintenance of systems of oppression. Inspired by Levinas's work, I made the case it is a moral obscenity to ignore the disproportionate impact climate change and biodiversity losses are having on minoritized people. A turn to the Other, must become an organizing principle in the method and matter of teaching. That said, vulnerability should be built into pedagogical praxis and extend to the more-than-human-world. EER, could open inquiries into communication practices that foster attunement and interconnectedness in face-to-face and online learning spaces. We need to better understand what gets in the way of attunement and what it looks like to name one's emotional experience, foster self-compassion and talk about the ethico-political relations enmeshed in the mis/recognition of climate affects in public education. Studies in EER could open space to reimagine the classroom or lecture hall as a relational home in which students can process bewildering and disorienting feelings about their changing environments and the impacts of those changes on Others.

Eco-pedagogies will empower young people with the knowledge they need to protect what remains of the rich diversity in the biological, cultural and social systems sustaining their precious lives. Over the next decade, teachers, school leaders and researchers, must move swiftly to work in solidarity with young people to mitigate the damage while deepening their collective understanding of the affective, cultural and mental registers of ecological degradation. Furthermore, EERs must identify and share the life-affirming strategies young people are already using to creatively cope with cascading waves of uncertainty. We need to do this difficult work even when it is hard to find our footing in a holding environment that is writhing out of kilter. Teaching in the Anthropocene obligates all of us to brace ourselves on unsteady ground. We urgently need classrooms in which teachers and students weave stories that acknowledge what we have lost, but more importantly, we need researchers, teachers and students, to write new stories to live by, stories to foster an intergenerational commitment to eco-attuned communities grounded in reciprocity and love.

The Anthropocene lays bare our debt to the Earth. Thousands of scientists and environmental activists from across the world tell us that we have but one generation to make massive changes to the way we live. When adults are dismissive about the severity of the climate crisis, it is an implicit denial of the value of a young person's life and the lives of young people yet to come. In the context of education, it is an abdication of *in loco parentis*, the fundamental obligation of educators to act in the best interest of a child when the child is in their care.

With that front of mind, I invite you to call in your colleagues, friends and family members, so we can collectively challenge narratives, policies and praxis in education that valorize autonomy, hyper-individualism and the commodification of other Beings in service of human ends. I hope this text contributes in some small way to a much larger project that centres mutual recognition as a fundamental aim of education. There is a shadow in the field cast by climate anxiety that makes it difficult to form new relational pathways among the more-than-human-world. However, monstrous feelings about the climate crisis can provoke us to stage our research avatars in new and unfamiliar scenes, thereby creating fresh interplays among memory, agency and imaginative reconstructions of self in relation to Others. To explore new trajectories and assemblages in educational research, breathe in, grab hands and tune into the Other. Let our professional lives be an answer to the question, *what does it mean to teach in a world that is prepared to go on without us?*

References

Abraham, N., & Torok, M. (1971). The topography of reality: Sketching a metapsychology of secrets. In Abraham, N., & Torok, M. (1994). *The shell and the kernel: Renewals of psychoanalysis* (pp. 157–164) (Vol. 1). (N. T. Rand, Trans.). University of Chicago.

Abraham, N., & Torok, M. (1972). Mourning or melancholia: Introjection versus incorporation. In Abraham, N., & Torok, M. (1994). *The shell and the kernel: Renewals of psychoanalysis* (pp. 125–137) (Vol. 1). (N. T. Rand, Trans.). University of Chicago.

Darabont, F. (Director). (1994). *The Shawshank Redemption. [Film]*. Columbia Pictures.

Ray Jacquette, S. (2021, March 21). Climate anxiety is an overwhelmingly white phenomenon. Scientific American. https://www.scientificamerican.com/article/the-unbearable-whiteness-of-climate-anxiety/

Appendix

The text block is a collection of either the headline or standfirst from each online news article published during the first two weeks of January, 2021 on three different news organizations' websites under the banner of climate change:

The Guardian (https://www.theguardian.com/environment/climate-change),
Aljazeera (https://www.aljazeera.com/tag/climate-change/) and
CBC News (https://www.cbc.ca/news/topic/Tag/Climate%20change).

I selected either the headline or the standfirst from each article based on which of the two best captured the content and context of each piece. The 61 headlines and standfirsts offer a snapshot in time of a mediated information assemblage that circulated online about climate change. By the numbers, CBC published one article, (#1), Aljazeera published 17 (#2–#18) and The Guardian published 43 articles (#19–#61). As you scan the text block, I encourage you to take notice of the words, phrases and images that give you pause, quicken your pulse, flood your eyes, flush your cheeks or make you want to click.

1. "Stop investing your money in climate failure", says one student. 2. "Dead heat": 2020 tied for warmest year on record, NASA finds. 3. UN: World Facing "catastrophic" temperature rise this century. 4. Signed, sealed, delivered: Sweden unveils Greta Thunberg stamp. 5. Iran's smog, blackouts made worst by power-sapping crypto mining. 6. What can corporations do to help save the ocean? 7. United States emissions fell by 10.3 per cent as the corona virus pandemic halted activity, a report from the Rhodium Group found. 8. Use the pandemic to protect forests, WWF urges consumers, politicians. 9. Early support for Mongolian herders as "extreme winter" looms. 10. Some HSBC shareholders urge bank to cut fossil-fuel lending. 11. One Planet Summit: World leaders to hold virtual climate

summit. 12. 2020 tied with 2016 as world's hottest year on record, EU says. 13. As mass livestock deaths threaten Mongolia's nomadic culture, we meet the herders trying to survive and those who've abandoned their traditional ways. 14. CO_2 levels seen more than 50% above preindustrial levels in 2021. 15. Plastic under scrutiny: Bank lending to industry faces opposition. 16. Finland's climate warriors. 17. Big oil sits out Trump's last-ditch Arctic drilling auction. 18. Why does Australia act as if it can ignore the climate crisis, and how long can it keep to this seemingly suicidal posture? 19. 2020 was hottest year on record by narrow margin, NASA says. 20. Revealed: Business secretary accepted donations from fossil fuel investors. 21. Western Australia LNG plant faces calls to shut down until faulty carbon capture system is fixed. 22. NGOs seek to convict French state of failing to tackle the climate crisis. 23. Climate crisis: record ocean heat in 2020 supercharged extreme weather. 24. Why are ocean warming records so important? 25. Top scientists warn of "ghastly future of mass extinction" and climate disruption. 26. BlackRock holds $85bn in coal despite pledge to sell fossil fuel shares. 27. National trust aims to save Yorkshire abbey from climate linked flooding. 28. US Greenhouse gas emissions fell 10% in 2020 as Covid curbed travel. 29. Dalai Lama says he "felt real hope" after hearing Greta Thunberg speak on climate crisis video. 30. Baby sharks emerge from egg cases earlier and weaker in oceans warmed by climate crisis. 31. It has been nearly four years since New Zealand experienced a month with below-average temperatures, researchers say. 32. More than 50 countries commit to protection of 30% of Earth's land and oceans. 33. Prince Charles urges businesses to sign Terra Carta pledge to put planet first. 34. Nigeria cattle crisis: How drought and urbanization led to deadly land grabs. 35. Shareholders push HSBC to cut exposure to fossil fuels. 36. Norway's electric car belies national reliance on fossil fuels. 37. 22 disasters, 262 dead, $95bn in damages: US saw record year for climate-driven catastrophes. 38. Alok Sharma to work full-time on COP-26 climate conference preparation. 39. Kenya faces $62bn bill to mitigate climate-linked hunger, drought and conflict. 40. Climate crisis: 2020 was joint hottest year ever recorded. 41. Last decade was hottest on record for Australia with temperature almost 1C above average. 42. Developing economies need fairer way to help them decarbonize. 43. UK's beef herd could be key to sustainable farming, says report. 44. Global heating could stabilize if net zero emissions achieved, scientists say. 45. Making waves: the hit Indian island radio station leading climate conversations. 46. Jenrick criticized over decision not to block new coal mine. 47. The barriers to a carbon fee and dividend policy. 48. Who should pick up the tab for the costs of climate change in north Queensland?

49. Severe climate driven loss of native mollusks reported off Israel's coast. 50. Revisited: What happens when the oceans heat up? 51. There's a simple way to green the economy- and it involves cash prizes for all. 52. UK urged to put Alok Sharma in full-time charge of Cop26 talks. 53. Australia's new climate pledge to UN criticized for not improving on 2030 target. 54. Australia inching forward to committing to net zero by 2050, top energy advisor says. 55. Climate crisis will cause falling humidity in global cities – study. 56. Greta Thunberg at 18: "I'm not telling anyone what to do". 57. UK car makers have three years to source local electric car batteries. 58. Jane Goodall: "Change is happening. There are many ways to start moving in the right way". 59. A wing and a prayer: How birds are coping with the climate crisis. 60. "The sea is rising; the climate is changing": The lessons learned from Mozambique's deadly cyclone. 61. Climate scientist says that another top 10 year is a "no shit, Sherlock" moment as temperatures across the country were 1.12C above average.

Index

Abraham, N. 7, 76, 88, 132–134, 164, 190
actants 9
activists 5, 18, 28, 60, 66, 78, 106, 184, 193
affectivity 60, 70, 95–96
Agrawal, A. A. 76, 104
Ahmed, S. 141
Albrecht, G. A. 75
Alexa 9, 140
ambivalence 89, 94, 102, 107, 138, 141, 182, 192; climate 6, 135, 139; field 3–6; leaders 11, 128, 159; monster 138–139; motivational interviewing 166–167; research 149, 156
American Psychiatric Association 155
animal dissections 110–111
Anthropocene 1–2
anthropocentrism 1–2, 10, 23, 43, 77, 101–102, 108–110, 153, 185
antisocial digital context 24–25
apocalyptic anxiety 66, 69, 148
asymmetries of power 3, 87
Archive 69–70, 140
artificial intelligence (AI) 69, 140
arts-based research 38, 47, 137–138, 191
attachment 7, 13, 34, 89–92, 97, 130, 138, 182–183
Attenborough, D. 69, 105
attunement: children's feelings 86, 177; definition 92, 153; eco-concerns 9, 11, 146; emotions 10, 28, 126, 189, 192; in-between spaces 33; learning environments 153, 177, 192; more-than-human-world 2, 17, 38, 180; research 149, 151, 153; telluric emotions 176; verbal and non-verbal language 177

bad object 64, 97, 128
Banksy 7, 81–82
Barthes, R. 39
Benjamin, J. 11, 65, 86, 89, 97, 129, 162, 167
Bill 64 168
binary thinking 151
biodiversity losses 36, 190–192
biomorphic powers 7, 137, 141, 146, 191
birth of eco-clinic 155
blue spaces 154
Boal, A. 138, 177
Boin, A. 2, 163
boomernomics 20
Bowlby, J. 89
Britzman, D. P. 176

Camera Lucida 39
carbon fair share 96, 113–114, 153
carp 38, 73–75
cascading crisis 28
cellphilm 140, 180
Change the Context 111–112
cheerleading 155
climate: café 183; conscious 5; crypt 77, 134, 137; dread 6, 64, 152, 190; goosebumps 58; science 4, 100 (*see also* IPCC); strike 3; swerve 107
climate psychology 85
Climate Psychology Alliance 75, 85
coercive dependency 162
colonization 19, 101, 141
compassionate anticipation 92–93
compassionate witnessing 10, 19, 75, 177–178, 190
community-mothers 152; We the Community Mothers in Education 152–153

complementary structure 162
Conceivable Futures 59
connectivity 122, 125, 132, 148
countertransference 12, 87–92
COVID-19 12, 18, 26, 42, 65, 113, 119–130, 155, 164, 178, 184, 189, 195
creeping crisis 2
crypt: climate 77, 134, 137; emotional 135, 192; leaky 134; psyche 76, 133, 138; social psyche 76
Cunsolo, A. 5, 58, 75, 115
cut to cure politics 124

dark arts of communication 168
dark tourism 25
datafication 26, 36, 64, 159, 169
Davenport, L. 70, 78, 141, 153
death anxiety 59, 69, 108
Decolonizing Water Project 178
Delta Marsh 73–74
denial 5, 10, 18, 60, 66, 68, 76, 78, 132, 134–135, 147, 149, 155, 160, 190, 193
denialism 68
Derrida, J. 132
dialogic fascination 151, 153–154, 181
digital storytelling 147
disaster porn 25
disavowal 1, 5–6, 67–69, 76–77, 82, 102–103, 133, 149, 164, 190
disenfranchised grief 6, 74, 77–78
Dismaland 7, 81–82
doom 66, 160
drama 111–112, 139, 141
dream of the burning child 136
dreams 16, 136, 139, 184

echolocation: digital 24–25; pedagogic 10; research 146
eco-anxiety 6, 37, 43, 63, 65–66, 70, 105, 135, 163, 189; research 149, 154–155; white privilege 18
ecocide 7, 62, 76, 141
eco-feminist 152
eco-grief 7, 73, 75–79, 81–82, 135, 180, 189
ecolinguistics 85, 101, 168–169
ecological debt 113
ecosophy 8–10, 138
ecosophical aims 17, 146
ecosophical orientation 6, 9–11, 67, 82, 85, 176, 178, 183

ego swelling 68
emergence 79, 87, 97
emotional labour 154
emotionally illegible 7, 78
enactments 9, 78–79, 101, 149, 150, 179
epistemic trust 95–96
ethnodrama 148
explicit mentalization 95

Farrell, A. 13, 15, 126, 137, 159
feeling-photography 2, 36, 38–41, 191
Ferenczi, S. 12–13, 87–89, 133
field of relations 11–12, 40–41, 44, 86–87, 127, 159, 165–166, 192
Flint Michigan 2, 17
Fonagy, P. 91, 95
Foote, E. 100
formative relationships 33, 90, 148, 181–182
Forum Theatre 138–139
Foucault, M. 154–155
Francis, Pope 27
freezing 61
Freud, S. 12, 79, 87–89, 108, 136, 176
Fridays for Future climate strikes 3, 10, 16

geese 33
Gentile, K. 13, 108
ghosts 65, 132; ancestors 76, 93; Anthropocene 7, 45, 76; epistemology 7, 132; language 150; question 132; reawaken 93–94; settler colonialism 94; trauma 164; unconscious 1, 94
Gillespie, S. 115
Government of Manitoba 80, 119, 122
Greene, M. 174–175
greenwashing 8–9, 76, 78, 190
Guattari, F. 8–9, 43, 137
guiding 155, 167, 172

Haraway, D. J. 38, 43, 67, 165
haunting vitality 12, 43
Her 140
Hickman, C. 5, 60, 93, 153
Hill Collins, P. 152
Hoggett, P. 27, 85
holding environment 64, 67, 79, 87, 90–91, 133, 162; classroom 6–7,

79, 91, 152, 177; compassionate witnessing 6, 177; dread 6, 64, 151; Earth 67, 93–94, 133, 192; leaky 94–95, 133–134, 162; research 64, 151–155; social media 12, 87, 94; Winnicott 12, 79, 87, 90–91, 177
hyper-object 36, 132, 184
hypnotism 177

implicit mentalization 95
incorporation 7, 76, 133–135, 190
Indigenous: disproportionate impact of climate change 100, 115, 125, 147; cosmologies 172, 178; eco-anxiety 66; knowledge keepers 184; leadership 18, 28; service minister 18; water injustice 18
in loco parentis 8, 135, 149, 172, 191, 193
insect apocalypse 22, 104
insect declines 103
intergenerational malpractice 5, 59
Intergovernmental Panel on Climate Change (IPCC) 4, 27–28, 68, 94, 134, 166
Intergovernmental Science Policy Platform on Biodiversity and Ecosystem Services 102
International Union of Geological Sciences 1
interpersonal psychology 89
intersecting crises 119, 126
introjection 74, 76, 82, 89, 133–135

Katniss 63
Kissi-Debrah, E. 114
Klein, N. 16, 176

Last Call for Sincere Liars 13
late stage capitalism 2–3, 114, 189
Latour, B. 116
Laudato Si 27
Leadership: compassionate 177; definition 159; ecosophical 159–164; ethic of care 122; feminist 152, 162; flips 160; queer 162; relational 129–130, 159; sympoietic pulse 165–166; vulnerability 122, 129, 130, 162
Lertzman, R. 60, 141, 153, 155
Levinas, E. 12, 65, 101, 126, 192
Lifton, R. J. 59, 61, 107, 112, 184
likes commodity exchange 96

mad max logic 2, 17–19
malignant normality 61, 112
Manhattan Institute of Psychoanalysis 61
Manitoba Education 4
Manitoba Flood of 1997 7, 79–80, 82
Matilda Effect 100
McKibben, B. 28
melancholy 13, 76, 93
mental health as plan 128–129
Mental Health Commission of Canada 145
mentalized affectivity 96
A Message from the Future 176
Mirror Instincts 92–93
mitákaye oyás' in 102
Mitchell, S. A. 11, 88–90
monster: ambivalence 138–139; aesthetic 42, 137; becoming 7, 137, 191; bring them closer 137, 140; inherit 137; interiority 7, 138–139; othering 141; play 7, 137, 139, 142; strength 7; warnings 137–138
monstrous: creation 11; feelings 5, 7, 11, 44, 63, 70, 137, 141, 163, 191–193
moody pessimism 7, 37, 41, 63, 82, 88
Mother Nature 105
motivational interviewing 166–167
Mni wiconi 2, 17–19
monarch butterfly 58, 104
Monsanto 105–107
Morton, T. 36, 44, 132, 184
mourning 7, 39, 41, 73–77, 82, 133–134
mutual recognition 10–12, 45, 67–68, 79, 87–90, 97, 126, 135, 140, 147, 150, 152, 156, 162–163, 176–177, 179, 190, 193
mystery 38, 60, 87, 141

Naess, A. 8–9, 146, 178, 185
narcissistic signals 68
narrative photography 2, 7, 36, 40–44, 48–56
natural language processing (NLP) 9–10
negation 68
Neo-Freudian 89
neoliberalism 20, 168–169; accountability 20, 22, 159, 168; austerity 20, 60, 123–124; competitive 124–125, 147, 159, 186, 191;

entitlements 96; fantasies and illusions 20, 85, 124; frames 11, 124, 159, 169; machinations 63; policies 159; politics 20; tropes 124
Norgaard, K. M. 61
not-me-possession 90

My Octopus Teacher 141
oikos 8
okay boomer 3, 151
Orange, D. 60, 65, 88, 93–94
Other: keeper 12, 14, 101; moral imperative 101; obligation 3, 6, 13, 28, 86, 96, 101, 125, 129–130, 135, 149, 172, 181, 191, 193; turn toward 101, 177, 192
othermothering 152
ouroboros-subjectivity 12, 101–102
outdoor classrooms 8, 160, 185

pedagogy of: affect 67; compassionate witnessing 177–178; deep ecology 178; dialogic fascination 181; feeling 179–180; interdependence 186; land-based learning 184; mitigation 182–183; multivocality 180–181; place 8, 172–173, 185; public 6, 77; self-analysis 182; social dreaming 183
pessimism 63
photographic sketches 38, 42
photovoice 147
piecemeal epistemology 37, 127, 147
politics of emotion 141
A Prairie Elegy for the Discerning Consumer 2, 43–44, 47–57, 177
preservative repression 7, 76
projection 6, 64, 68, 90, 122, 190
psychic numbing 61
punctum 39
purr-words 63

registers 13, 41, 120, 137–138, 147, 156; affective 3, 6, 13, 145, 149, 152, 156, 161, 163; cultural 166; ecological 9–11, 120, 146, 163, 168; emotional 5, 75, 126, 191; ideological 138; mental 43, 192; social 43
regularization of happy affects 62
relational configurations 90
relational interviewing 167
relationalists 96

Replika 140
research creation 47, 191–192
rhythmic affect 91
rhythmic thirdness 97
righting reflex 167
Ringstrom, P. A. 11, 90
Roy, A. 128

sacrifice zones 85, 101, 126
Sangho, B. 139
Scholossman, M. 7, 77, 82, 92
Searles, H. F. 66, 69, 153
selfies 42, 186
Self-portrait as a Drowned Man 40
sensory-struck 6, 38, 87
shame 60, 65, 108, 110, 124, 141, 159
Shawshank Redemption 189
Shiva, V. 106
silent spring 104
siren song of grievance 21
Siri 140
sixth mass extinction 2, 27, 76, 103
slippery talk 168
social dreaming 183–184
social-emotional necrosis 3, 47
social media: algorithms 25; anti-social 94–95; anxiety & distress 26, 95, 122; avatars 24; images 5, 42, 59, 70; discourse 2, 17, 42; game within a game 24; habitus 3, 5, 95; identity; likes as commodity 95–96; memes 25; misinformation 60; platforms 24–25; use 95
Solastalgia 75
somatic: awareness 162; echoes 132; experience 136; poetics 183
Sontag, S. 81
Specter of Marx 132
staged photography 40–41, 48–56
Stern, D. B. 11, 86, 186
Stibbe, A. 63, 101–102, 168
Stoknes, P. E. 122, 160
Stolorow, R. D. 67, 76, 128
stories we live by 82, 101, 105, 108
storytelling 44, 47, 70, 137, 141, 147
suffering 45, 106, 111, 125, 191; animals 112, 184; Other 28, 60, 122, 126, 147, 156, 164, 167; stranger 25, 65, 75, 147, 153, 190; trying it on 25; young people 67
Sullivan, H. S. 89

Suzuki, D. 27, 102
sympoiesis 166

tattoo 75–76, 104
teacher education 8, 43, 80, 88, 172–187
teacher thought bubble 181–182
technology: consequences 1, 66, 77, 178; fetishizing 76, 105; Pope Francis 27; professional development 127; protective 79, 125; saviour 105
telluric emotions 20, 60–64, 75, 139, 149–150
terra cremata 64
think disobediently 47, 81, 191
Thunberg, G. 10, 78, 134–135, 194, 195–196
Torok, M. 7, 76, 128, 132–134, 164, 190
transitional objects 90
trauma: capacity building 81, 128, 163; climate/eco 7, 20, 67, 70, 76, 85, 132, 156, 189; effects 65, 69–70, 109–110, 128, 163–164, 190; events 61, 133, 177; exposure response 20, 122, 126, 163–165; hidden/suppressed 7, 77, 133, 190; higher-order 164; informed practice 128; intersubjective 19, 64, 67, 128, 163; meta psychological 134; post-traumatic stress disorder 19, 80; research 20, 151, 163–164; stewardship 155–156; unprocessed 13; vulnerability 163
traumatology 164

uchronic stories 165
unbidden 8, 11, 43, 87, 93, 148, 172, 174
unburied crimes 94
uncertainty 37, 96, 121, 148, 164, 175, 177, 192
ungrading 186
United Nations Environment Programme 1, 112–114

van Dernoot Lipsky, L. 155
visuality 2, 7, 11, 17, 33, 36, 38, 42–43, 47, 126, 132, 148, 166, 174, 184
visual research 44, 147
vulnerability 6, 12, 64, 76, 122, 127, 129–130, 147, 163, 177, 192

Wagamese, R. 38
Wallin, J. 63
water futures 18–19
weather talk 58, 73, 132
Weintrobe, S. 60, 68, 97, 151, 153
wildfires 17, 62, 78, 119, 183–184
wilding field experience 185
windshield phenomenon 104
Winnicott, D. W. 79, 90–91
withholding witness 65, 150
witnessing professional 61
Woodbury, Z. 164
word boxing 124

Xerces Blue butterfly 6

Yu, T. 139